CITY OF THE
STARGAZERS

CITY OF THE STARGAZERS

By Kenneth Heuer

Illustrated with photographs and drawings

CHARLES SCRIBNER'S SONS ★ NEW YORK

THIS BOOK PUBLISHED SIMULTANEOUSLY IN
THE UNITED STATES OF AMERICA AND IN CANADA—
COPYRIGHT UNDER THE BERNE CONVENTION

A—9.72(H)

PRINTED IN THE UNITED STATES OF AMERICA

Library of Congress Catalog Card Number 72–1175
SBN 684–12937–X (cloth)

FRONTIS. *Detail from the Alexander Sarcophagus.* The Alexander Sarcophagus was found in a vault at Sidon, a powerful city-state of ancient Phoenicia. The marble tomb is miraculously well preserved and extols the fame of the Macedonian king who disseminated Hellenic culture throughout the eastern Mediterranean and western Asia. One side of the sarcophagus represents Alexander at the battle of Issus (October 333 B.C.) in which he triumphed over Darius and the Persians, and the other shows him conducting a lion hunt with a party of Persian guests. This fine early Hellenistic work of art dates from the late fourth century B.C.

To ☆ EMILY FAITH CAIGAN

The cities which have remained great and glorious over long periods of time are those with a rich variety of population, economic enterprise, and social functions. Diversity endows them with the resilience required for surviving the upheavals of history and for maintaining identity in the midst of endless changes.

RENÉ DUBOS: *A God Within*

Contents

CONTENTS

Illustrations

ILLUSTRATIONS

ILLUSTRATIONS

Preface

If you look at a map of the world, you will see that Alexandria is situated on the Mediterranean Sea at 31°12' north latitude and 29°15' east longitude. A chief seaport, this city is located 129 miles by rail northwest of Cairo, the present capital of Egypt. The Canopic mouth of the Nile that existed long ago was 12 miles east; it is now dry.

Alexandria is an ancient city. For over 1000 years from its foundation it was the capital of the country. It is to this city that I refer in *City of the Stargazers.*

The stargazers of the title are the astronomers who lived and labored in ancient Alexandria. While that city was ruled by the Greeks and the Romans, a pagan school of science, literature, and philosophy flourished there. So Alexandria could as truly be called "city of the mathematicians" or "city of the grammarians" or "city of the philosophers." But I have chosen to tell primarily the story of the distinguished men of Alexandria who looked upward and were led from this world to the study of other worlds in space.

These men who centuries ago tried to explain the heavens were more than astronomers. Claudius Ptolemy, for example, was a geographer and mathematician as well. Thus any specialist (and I am one) who writes about these scientists is in danger of giving an unbalanced picture. Here the emphasis is on these men as astronomers, and most space is given to their astronomical work.

City of the Stargazers is an attempt to take the facts and figures of science out of the vacuum in which they are usually given and to present them against the age in which the discoveries were made and in the context of the lives and other accomplishments of the men who made them.

Many have heard that Aristarchus originated the sun-centered hy-

15

pothesis in astronomy, that Eratosthenes measured the circumference of the earth, that Hipparchus discovered the precession of the equinoxes, that Ptolemy centered the solar system upon the earth. But who were these men and how did they come to make their discoveries? How were they shaped by their surroundings and by the events of their times?

As the stargazers' story unfolds, so does the history of Alexandria, a city whose name is familiar but about which most people know very little —in short, a city lost in time. As the reader progresses through these pages, he meets monarchs and murderers, poets and patriarchs, ordinary and extraordinary people, and the topography of the city falls into place piece by piece until it is complete.

Descriptions in literature, some papyri, coins, contemporary and later painting, sculpture, and minor artistic work reveal a number of the striking aspects of Alexandria and are the basis of the illustrations in the book.

My aim in this book is not to make a scholarly contribution but rather to tell about a fascinating period of history. In this endeavor I have been helped by the following scholars who read all or part of the manuscript and made suggestions: Dr. I. Bernard Cohen, Professor of the History of Science at Harvard University; Dr. Louis Cohn-Haft, Professor of History at Smith College; Dr. Walter A. Fairservis, Jr., Professor and Chairman of the Department of Anthropology and Sociology at Vassar College; and Dr. Owen Gingerich, Professor of Astronomy and of the History of Science at Harvard University. To all of them I wish to express my gratitude.

Ancient Alexandria is gone; the lights of that magnificent city are extinguished forever. But in the darkness behind us are bountiful kings, the Ptolemies, and with them the great scientists and thinkers and men of letters at work in the famous museum and library. The story begins with the founding of the city.

16

Chronology

The story of ancient Alexandria has so many themes that a departure from a strict time sequence is sometimes necessary. The following chronological table includes important dates relating to the history of Alexandria, from its founding to the destruction of its walls. For the reader's convenience, it is placed before the text.

In the field of chronology, the first scientific attempt to fix the dates of political and literary history was made in Alexandria more than 2,000 years ago. This pioneer effort was made by Eratosthenes, one of the main characters in this narrative, who also compiled a list of Olympic victors.

332 B.C.	Alexandria is founded by Alexander the Great.
Late 4th cent. B.C.	Dinocrates designs the city of Alexandria.
332–331 B.C.	Cleomenes of Naucratis is appointed financial manager of Egypt and administrative chief of the eastern Delta district by Alexander.
323 B.C.	Alexander dies of fever in Babylon.
323–306 B.C.	Ptolemy, one of Alexander's generals, is satrap of Egypt.
322–321 B.C.	Cleomenes becomes hyparchus of Egypt under Ptolemy.
307 B.C.	Demetrius Phalereus takes refuge in Alexandria.

17

CHRONOLOGY

305 B.C.	Ptolemy assumes title of king.
300–294 B.C.	Strato of Lampsacus, "the physicist," is tutor to Ptolemy II.
About 300 B.C.	Euclid, author of the *Elements,* teaches in Alexandria.
	Herophilus founds a school of anatomy at Alexandria.
	Erasistratus is active as a physician and anatomist.

First half 3rd cent. B.C.	Theocritus creates pastoral poetry as a form.
285 B.C.	Ptolemy I resigns in favor of his son Ptolemy II.
285–246 B.C.	Ptolemy II is king of Egypt.
About 285–283 B.C.	Lycophron is entrusted with the preliminary sorting out of the comedies collected for the library.
About 284 B.C.	Zenodotus becomes the first head of the library at Alexandria.
281 B.C.	Aristarchus of Samos observes the summer solstice at Alexandria.
About 270 B.C.	The architect Sostratus builds the Pharos (lighthouse).
About 264 B.C.	Callimachus of Cyrene is active as poet and scholar.
About 260 B.C.	Apollonius of Rhodes succeeds Zenodotus as librarian.
Middle 3rd cent. B.C.	Archimedes studies with the successors of Euclid at Alexandria.

18

| Second half 3rd cent. B.C. | Apollonius of Perga, mathematician known especially for treatise on conic sections, flourishes. |

| 246–221 B.C. | Ptolemy III, son of Ptolemy II, is king of Egypt. |

| About 245 B.C. | Conon of Samos names the star cluster Coma Berenices. |

| About 235 B.C. | Eratosthenes succeeds Apollonius of Rhodes as librarian. |

| 221–203 B.C. | Ptolemy IV, son of Ptolemy III, is king of Egypt. |

| About 194 B.C. | Aristophanes of Byzantium succeeds Eratosthenes as head of the library. |

| About 153 B.C. | Aristarchus of Samothrace succeeds Aristophanes of Byzantium as head of the library. |

| About 130 B.C. | Hipparchus discovers precession of the equinoxes. |

| 80 B.C. | Ptolemy X (or XII), last of the legitimate male line of Ptolemies, is killed by the enraged populace. Alexandria passes formally under Roman jurisdiction. |

| 48 B.C. | Gaius Julius Caesar sets fire to the Egyptian fleet in the harbor of Alexandria. |

| 40 B.C. | Marcus Antonius presents the library of Pergamum to Cleopatra. |

| 30 B.C. | Cleopatra kills herself and the dynasty of the Ptolemies ends. Alexandria becomes the capital of a Roman province. |

| Second half 1st cent. A.D. | Heron of Alexandria writes numerous works in mathematics, physics, and mechanics. |

19

CHRONOLOGY

A.D. 127–141	Claudius Ptolemy makes observations of the heavens.
A.D. 157	Galen starts practising medicine in Pergamum.
First half 3rd cent. A.D.	Ammonius Saccas, one of the founders of Neoplatonism, teaches at Alexandria.
A.D. 215	The Alexandrian youth are murdered by order of Caracalla.
A.D. 244	Plotinus, chief exponent of Neoplatonism, begins lectures on philosophy in Rome.
About 250 A.D.	Diophantus, reputed inventor of algebra, flourishes in Alexandria.
A.D. 273	Aurelian puts down a revolt in Alexandria.
A.D. 296	Diocletian subdues a fresh revolt in Egypt, after subjecting Alexandria to a siege of eight months.
A.D. 323	Constantine I chooses Byzantium as his new capital.
A.D. 325	Constantine calls the Council of Nicaea at which the Nicene Creed is adopted.
A.D. 328–373	Athanasius is patriarch of Alexandria.
A.D. 330	Byzantium, chosen by Constantine as the site of his new capital of the Eastern Roman Empire, is renamed Constantinople.
Second half 4th cent. A.D.	Oribasius is personal physician of Emperor Julian and compiles a medical encyclopedia.

20

Second half 4th cent. A.D.	Theon of Alexandria edits Euclid's *Elements* and writes a commentary on the *Almagest.*
A.D. 385–412	Theophilos is bishop of Alexandria.
A.D. 400	Hypatia becomes head of the Neoplatonic school in Alexandria.
About 415 A.D.	Hypatia is murdered by a mob incited by Cyril, archbishop of Alexandria.
A.D. 428	Nestorius is appointed patriarch of Constantinople.
A.D. 431	Nestorius is deposed for heresy and banished to the Libyan desert.
About 450 A.D.	Proclus, last of the great teachers of Neoplatonism, teaches in Athens.
5th cent. A.D.	The Brucheum and Jewish quarters are abandoned.
A.D. 529	Justinian I forbids the study of all "heathen learning" in Athens; the school of Athens dies.
A.D. 616	Alexandria is taken by Khosrau II, king of Persia.
A.D. 641	Alexandria is surrendered to Amru by Cyrus, patriarch of Alexandria.
A.D. 645	Manuel, commander of the Imperial forces, attempts to regain Alexandria.
A.D. 646	Amru conquers and destroys Alexandria.

CITY OF THE STARGAZERS

PART ONE ✵✵✵ THE RISE

OF THE CITY

1 ✫ A City Is Founded

Alexander the Great wept when he learned that there is a countless number of worlds. Anaxarchus, a philosopher of Abdera who accompanied him into Asia 334 years before the birth of Christ, told him this. When Alexander's friends asked him if he had been injured, according to Plutarch in the *Moralia*, the conqueror replied: "Don't you think it is a matter worth crying about that when there is such a vast number of worlds we have not yet conquered one?"

Alexander, king of Macedon, conquered Asia as far as western India, but while he was making plans to subdue other countries, the thirty-three-year-old leader was stricken with a fever and died in 323 B.C. After his death his companion Anaxarchus was shipwrecked and fell into the power of the Cypriot prince Nicocreon. The philosopher had offended the ruler, who had him pounded to death in a stone mortar.

When Alexander invaded Africa, Egypt hailed him as its ruler without a struggle. And while the conqueror's troops were stationed at the mouth of the Nile, he began the construction of a beautiful city, Alexandria. In every respect, Alexandria was admirably suited to become the new center of the world's activity and thought. Its geographical position brought it into commercial contact with all

the nations lying around the Mediterranean. At the same time, its situation made it a vital communicating link with Eastern wealth and civilization. These tremendous natural advantages were increased to a great extent by the care of Egypt's sovereigns, the Ptolemies.

Alexander founded a great number of cities. They were located with such wisdom that many of them became flourishing centers of industry and outposts of Greek civilization. From Greece to Asia Minor and Egypt and even beyond came merchants, scholars, artisans, and ex-soldiers. Naturally Greek culture could not expand in this way without being changed to some degree by Oriental civilization. A real union between the Greek and the Oriental worlds took place, and this fusion gave rise to a period which became notable for achievements in science, art, and philosophy. This later age of Greek culture is known as the Hellenistic (or Greek-like) period, as distinguished from the earlier Hellenic (or classical Greek) period.

Head of Alexander the Great. Unearthed at Alexandria, the marble head dates from the second century B.C. Its delicacy of feature and the somewhat oversentimentalized set of the head make it typical of the high romantic phase of Hellenistic sculpture.

2 ✲ A Dynasty Begins

Ptolemy founded a dynasty, a succession of kings of the same family who ruled in Egypt for almost 300 years. He had been one of Alexander's most trusted generals and among his seven "bodyguards." He had played an important part in the later campaigns of Alexander in Afghanistan and India and took Egypt as his share of Alexander's empire when it was divided after the brilliant military leader's death. In the breakup of the empire, two other kingdoms were also created: Macedonia and Greece ruled by the Antigonids and Western Asia and Iran ruled by the Seleucids. But these three kingdoms emerged only after Alexander's generals engaged in a bewildering series of palace intrigues and wars, in the course of which Alexander's half-brother, infant son, and mother were murdered.

Ptolemy made Alexandria his capital and the foremost city in the world. He had a great psychological advantage over the other Hellenistic rulers, for it is told that he had managed to obtain Alexander's body. In life, Alexander had an athletic frame, though he was not unusually tall. His complexion was ruddy, his eyes were liquid and melting, though in times of anger they changed and burned brightly. The hair which stood up over his forehead gave the

Alexander wearing an elephant's skin headdress. The conqueror is shown on the obverse (the side of a coin bearing the principal image or inscription) of a silver tetradrachma, issued in Egypt during the reign of Ptolemy I. It suggested Alexander's conquest of India and recalled Egypt's possession of the hero's body. For to the ancients the elephant symbolized royalty and also deification. (This and other coins reproduced in the text are enlarged for clarity; they are shown in their actual sizes in the Chronology.)

suggestion of a lion, and he carried his head in a slanting direction. At least this is the notion of the personal appearance of Alexander that one gets from the literature and surviving monuments. The embalmed body was wrapped in linen and bound in plates of gold so formed as to preserve the beautiful contours of the hero. This Ptolemy placed in a magnificent tomb of rare Greek and Egyptian marbles in the heart of Alexandria.

Ptolemy, who became satrap, or governor, of Egypt in the late summer of 323 B.C. assumed the title of king in 305. Shrewd and cautious, he had a compact and well-ordered kingdom to show after many years of wars (he made himself master of Jerusalem by attacking the city on the Sabbath day). His good nature and generosity attracted the floating class of Macedonians and Greeks to his service, and he did not neglect to obtain the friendship of the native Egyptians.

Called Ptolemy Soter (savior), Ptolemy I was a ready patron of learning. From various parts of Greece he drew around him a circle of men distinguished in literature, philosophy, and science. These men were given every aid for the execution of their researches. Ptolemy's friend Demetrius of Phaleron (a coastal suburb of Athens) inspired him to lay the foundations of the Alexandrian library. This Athenian writer and statesman, who had been absolute governor at Athens, had fled to Alexandria to escape execution when democratic government was restored in 307 B.C.

Ptolemy himself wrote one of the best of the histories of Alexander the Great, drawing upon Alexander's Journal and other official material, as well as his own memory of a great many important things. Although it was more likely to have been a highly personal memoir, he may have used the library when he composed his work; he would have been one of the first historians to have done so.

32

Marble mask of Ptolemy I. In Egypt, marble, being imported and expensive, was used in making only certain parts of a statue. This mask of a portrait statue found in that country closely resembles certain coins of Ptolemy I which depict him in the prime of life. Showing none of the idealization conspicuous in the portraits of Alexander, those of Ptolemy I on the coins and in sculpture are always handled realistically. The main traits of his character are emphasized: iron power of will and determination guided by practical wisdom and intelligence.

For the convenience of his scholars, Ptolemy also built a museum. This was in the first place an institute for research in science. It probably contained dormitories for the scientists and their assistants and pupils, assembly halls, laboratories and an observatory, botanical and zoological gardens, and roofed colonnades for open-air study or debate. The museum did not include all these features in the beginning, of course; it grew with the passage of time.

Toward the end of his life, in 285, his youngest son was elected joint ruler with him. He lived two years longer, dying in 283.

Alexander the Great as Hercules. The delicate little drawing, in pen and brown ink on paper with a brown wash over it, is one of six similar sketches made by Peter Paul Rubens (1577–1640) from ancient coins and gems. It shows the careful study the Flemish painter gave to these objects of antiquity which he often mentioned in his correspondence.

3 ✦ *A Father's Task Is Completed*

Ptolemy II was of a delicate constitution. No Macedonian warrior chief of the old style, he enjoyed a relatively peaceful reign and maintained a magnificent court at Alexandria, where elaborate religious rituals and pomp were the order of the day.

The Ptolemaic kings adopted the customs of the pharaohs, the earlier sovereigns of Egypt, including marriage between royal brothers and sisters; Ptolemy II took as his second wife his own sister Arsinoë II. Egyptians had believed in the divine nature of their rulers. Therefore, it was natural enough for the Ptolemaic kings to invest themselves with divinity, and the second Ptolemy deified his parents and, after her death, his sister-wife, the latter as Philadelphus. The surname Philadelphus (brother loving) belongs properly to Arsinoë only, but it was used in later generations to distinguish Ptolemy II himself.

Moreover, each Egyptian dynasty had either given a new emphasis to one of the existing gods or introduced a new god. In the same spirit, the Ptolemies put Serapis in a conspicuous position before the public. It is not surprising that this deity combined the attributes of many of the mighty Hellenic gods with some of the characteristics of Osiris, the great Egyptian god of the underworld and judge of the dead.

Head of Ptolemy I and Berenice I. The king and queen are depicted on the reverse (the side of a coin opposite to the one bearing the principal image) of a gold octadrachma, issued under Ptolemy II or Ptolemy III. The place of coinage is uncertain but may have been Alexandria. Berenice I was the wife and half-sister of Ptolemy I and the mother of Ptolemy II.

Heads of Ptolemy II and Arsinoë II. This royal couple is represented on the obverse of the coin portraying Ptolemy I and Berenice I. The portrait of Ptolemy II well illustrates his personality: the softening of fiber which became more pronounced in several later kings already showed itself in the son of the tough old Macedonian general.

Ptolemy II's chief care was directed to the internal administration of his kingdom and to the patronage of literature and science. The museum became the resort and home of all the most eminent men of learning of his time, and in the library attached to it were collected the treasures of ancient knowledge, including Aristotle's library. Ptolemy also had the strange beasts of distant lands sent to the capital of Egypt. He is said to have completed his father's task; the enthusiastic and vigorous activity along literary and scientific lines was chiefly effected by these two kings.

There is a tradition that the Holy Scriptures of the Jews were translated into Greek at Ptolemy's command, but this is not historical. The earliest Greek translation of the Old Testament is known as the Septuagint, a name taken from the Latin word for seventy. It is derived from the legend that the translation was made by seventy or seventy-two scholars; supposedly Ptolemy ordered seventy-two Palestinian translators to come to Alexandria to translate the Hebrew books of the Law. A later legend maintains that these men were given separate rooms and were not permitted to talk to one another; yet their translations were exactly the same—proof that they were inspired by God!

The account of the translation has an element of truth. The Pentateuch, the first five books of the Bible taken collectively, was translated into Greek in Alexandria during the third century before Christ. However, it probably was translated to meet the needs of the large Jewish community there, which had all but lost its knowledge of Hebrew, instead of at the command of Ptolemy. However, Ptolemy was a ruler of liberal literary tastes, and he may well have promoted a project that not only would have appealed to his curiosity but would encourage the use of the Greek tongue among the Jewish population of Alexandria.

The actual number of translators employed in this task and who they were is a mystery. In the following centuries, the other books of the Old Testament were gradually translated. The Septuagint was virtually completed before the beginning of the Christian Era. Its value has been immeasurable. It made the Bible widely easy of access. It served as the authoritative Old Testament text for early Christianity. It was the basis of other translations of the Bible.

The material and literary brilliance of the Alexandrian court was at its height under Ptolemy II, who ruled from 285 to 246 B.C. His court has been appropriately compared with the Versailles of Louis XIV. It was splendid and influential, dissipated, artificial, and intellectual.

Silver armlet. The metal industry thrived in Hellenistic Egypt. As evidence of this, gold and silver plate, cult utensils, and jewels of the early third century B.C. have been found in abundance. During Ptolemy II's reign, hundreds if not thousands of gold and silver vessels and other objects were displayed at the pomps and banquets this king organized. The silver armlet in the form of coiling snakes shown is said to be from Balamun in the Nile delta. Snake bracelets are common in Greek art.

39

4 ✰ The Thousand-Year School

The eldest son of Ptolemy II was the next king of Egypt. Soon after he succeeded to the throne, he embarked upon a career of conquest. When he returned to Egypt, he brought with him the spoils of war. This immense plunder is said to have included all the statues of the Egyptian gods and goddesses which had been carried off to Babylon by Cambyses, Cyrus's son and the second king of Persia, who had conquered Egypt three centuries before. Ptolemy III restored the statues to the temples of the respective deities, thus obtaining the title Euergetes (benefactor).

Ptolemy Euergetes is also celebrated for his patronage of literature and science. Under him the library grew larger. He obtained the official Athenian copies of the works of the dramatists, and he had all books brought into Egypt by foreigners seized for the benefit of the library. The owners had to be content with receiving copies of them in exchange.

Ptolemy III reigned from 246–221 B.C. His eldest son and successor, Ptolemy IV, called Ptolemy Philopater (loving his father), did not inherit the abilities or virtues of his parent. His reign was the beginning of the decline of the Egyptian kingdom.

The intellectual movement begun in the manner described

Head of Ptolemy III. The king is depicted on the obverse of a gold octodrachma of Alexandrian coinage; it was struck after his death. The portrait clearly shows the inherited tendency of the kings of this house to grow fat in later life. Nevertheless, the king's appearance is rendered impressive by his fine profile, his royal air, a crown of rays adorning his head, and the trident of the sea god Poseidon at his shoulder.

Bust of Ptolemy IV. Ptolemy IV (shown here on the obverse of an octodrachma probably coined in Egypt after the reign of this monarch) was a weak ruler under the influence of court favorites. He is depicted on the coin as fat but still quite good-looking, a self-indulgent and self-satisfied man with sideburns.

lasted for nearly a thousand years—305 B.C. through A.D. 642. This period falls into two parts. The first was from the foundation of the Ptolemaic dynasty till Alexandria was brought under Roman dominion (305–30 B.C.). The second, 30 B.C. to A.D. 642, ended when Alexandria was conquered by the Arabs. The intellectual activity in the first period was literary and scientific. This was especially apparent under the early Ptolemies. Alexandria was at that time among the few homes of pure literature in the world. During the century and a half that preceded the Christian era, the school started to crumble and lose its distinctive character. This was caused partly by the state of government under some of the later Ptolemies, to say nothing of the pressure of Roman arms, and in some degree by the emergence of new literary circles on the island of Rhodes, in Syria, and elsewhere.

5 ✧ *The Problem with Homer*

The librarians of Alexandria had a difficult task. The first librarian, Zenodotus of Ephesus, had to identify the books (an ancient "book" was a convenient-size papyrus roll) and assemble those that belonged together. A notable instance was the books of the *Iliad* and *Odyssey*, which had to be identified, classified, and then edited. Zenodotus's critical revision of the *Iliad* and *Odyssey*, in which for the first time each of these epics was divided into twenty-four books, was the first scientific attempt to get back to Homer's original text by the comparison of different manuscripts. It should be noted that the division into twenty-four books is more a convenience for readers and librarians as well as scholars than a work of critical scholarship. And the earliest effort to produce a correct text of Homer, while undoubtedly primitive by Alexandrian standards, occurred in Athens in the sixth century B.C. under the patronage of the tyrant Pisistratus and was a famous event.

The process used with the works of Homer had to be followed for all the books in the Alexandrian library. It was necessary to establish the text of each author and to set standards by which to determine the correctness of a judgment. That is to say, Zenodotus and the librarians who came after him were also philologists.

44

Philology is the study of literature in a wide sense, and Eratosthenes of Cyrene, who became the librarian in about 235 B.C., was the first man to call himself a philologist. His most important work of scholarship was *On Ancient Comedy*. This was in twelve books. There he discussed literary, historical, and other matters and dealt with problems of the authorship and production of plays. Eratosthenes also laid the foundation of scientific chronology; he tried to fix the dates of the main literary and political events from the conquest of Troy.

The identification of texts and their establishment led to the development of every branch of philology. A knowledge of grammar, for example, was required to determine the sense of a text with certainty. Moreover, grammar became necessary for the teaching of Greek to foreigners in a city like Alexandria where several tongues were spoken.

Aristophanes of Byzantium and Aristarchus of Samothrace were, strictly speaking, the first grammarians. Aristophanes, a scholar of wide learning, edited the works of a number of important writers. His edition of the *Iliad* and *Odyssey* was an unmistakable improvement on those of Zenodotus and of Rhianus of Crete; the latter was a contemporary of Eratosthenes who started life as a slave and custodian of a wrestling school and, after an education delayed beyond the usual time, became famous as a poet and Homeric scholar at Alexandria. Aristophanes also wrote a grammatical work in which he tried to define the rules of Greek declension.

Aristarchus, who founded a grammatical and critical school at Alexandria, was probably the greatest critic of antiquity. His works also include an edition of Homer. There were, as already implied, textual problems in the *Iliad* and *Odyssey*. The text had been rendered incorrect by changes or errors. Judgments had been made

Early Hellenistic vase. This blue-glazed vase, with base and mouth restored, was made in Ptolemaic Egypt and is dated 200–100 B.C. It is embellished with six friezes with figures and ornaments in bluish green. On the neck, in the spaces between clusters of papyrus stalks, there are superhuman figures, probably representing the Nile god. The next frieze containing figures shows fish and birds and between them flowers and fruit. The frieze in which real and fantastic animals are depicted is the most from incomplete evidence. Unauthorized material had been inserted. Aristarchus was more careful in the treatment of these problems than the Alexandrian scholars who had gone before him. He endeavored to solve them by seeking out the manuscripts which seemed closest to the original words of the poet, as determined by careful study of Homeric language and meter, by Aristarchus's fine literary sense, and by his practice of interpreting a poet by the poet's own use of words or forms, among other means. In his old age Aristarchus left Alexandria for Cyprus. There, at the age of 72, he starved himself to death because he was suffering from an incurable disease—dropsy.

The merit of these men, all of the third and second centuries B.C., is to have collected, edited, and preserved the existing monuments of Greek literature.

The poets of the Alexandrian school, from the same period, included Lycophron, a native of Chalcis in Euobea; Callimachus, born in Cyrene; Apollonius of Rhodes; and Theocritus, a native of Syracuse. At Alexandria Ptolemy II placed Lycophron in charge of the preliminary sorting out of the comedies collected for the library. He was the author of a poem on the fall of Troy, called the *Alexandra*, which still exists and is famous for its obscurity. The poet was ultimately killed by an arrow.

Callimachus immigrated to Alexandria early in life and taught school in its suburb Eleusis. His longest and most famous work was the *Aetia*. In the prologue of this narrative elegy, Callimachus described himself as being carried in a dream from Libya to Helicon, where he questioned the Muses about ritual, myth, and history. How far this framework was kept in the poem is not known. Callimachus wrote numerous poems. His boldness in experiment and his ability to write in different styles displeased those of his time

Mosaic celebrating Ptolemaic maritime supremacy. The tepidarium of a bath in a villa near Leptis Magna, an ancient seaport in Roman Africa, was decorated with a mosaic of four panels. All four glorify the blessings bestowed on mankind by water; the one reproduced here shows the sea as carrier of maritime trade. To the extreme right, there is a harbor and behind it a palatial building. Nude Cupid-like children, often used in decorative painting and sculpture, are hurrying to the harbor, some flying, others riding dolphins, two sailing in a warship, all bearing flowers and fruit. Although of Roman origin, the panels of the mosaic express the Hellenistic Alexandrian spirit, the same that is found in the poetry of the period, and each appears to deal with Alexandria and Egypt. This one extols the maritime supremacy of the Ptolemies (the warship), devoted to the service of Ptolemaic commerce; the combination provided wealth, merriment, and abundance for Alexandria, Egypt, and the king himself (symbolized by the palace in the harbor).

48

opposed to change or innovation, but they show sufficient grounds for the high regard in which he was held in later ages.

Apollonius has generally been called "the Rhodian" because he finally went to live in Rhodes. His most important work was the *Argonautica*, an epic based on the legend of the Argonauts, the heroes who sailed with Jason in the ship *Argo* in quest of the Golden Fleece.

Theocritus, the creator of pastoral poetry, visited Alexandria during the end of the reign of Ptolemy I. For his first efforts, he obtained the patronage of Ptolemy II, who ruled with his father at the time, and in whose praise Theocritus wrote a number of his poems or idylls.

6 ☆☆ *The Cadaver Is Questioned*

Among those who pursued medicine, mathematics, physics, and astronomy in Alexandria from the third century B.C. to the third century A.D. was the anatomist Herophilus of Chalcedon, who founded a school and was one of the first to supervise examinations done after death. (The allegation has been made that the Alexandrian anatomists dissected the bodies of living men, in order to have a better understanding of the functioning of the organs, but there is no compelling evidence to accept or reject this.) In this way, he made as many discoveries as would an explorer who was the first to visit an unknown planet. He carried out this ambitious program of anatomical research systematically under the rule of Ptolemy I. His slightly younger contemporary, Erasistratus of Ceos, continued the survey of the human body but paid more attention to physiology. It is said that Erasistratus had an incurable ulcer on his foot and that he committed suicide by drinking hemlock.

The fact that both anatomy and grammar were developed in the Hellenistic age is a natural circumstance. Both were the outcome of the same analytical and scientific mentality, applied in the case of anatomy to man's body and in the case of grammar to his language.

The mathematician Euclid, who lived in Alexandria in the

Euclid looking on mathematics bare. One of the earliest and most famous men of science associated with the new metropolis of Alexandria was Euclid. His name and his main work are familiar, but there is little information about the man himself. What is known is inferred and of late publication, such as this picture by André Thevet which appeared in *Portraits and Lives of Illustrious Men* published in Paris in the late six-teenth century. Such ignorance is not unusual where the great scientists of the past are concerned. Unlike the monarchs, they were not considered important enough to be represented on coins of the realm or sculptured in marble. Like a poet, a mathematician does not need close collaborators. Here Euclid is seen drawing geometrical figures, working quietly by himself, probably in his own home.

time of the first Ptolemy, also founded a school there. Several of his works are still in existence. The most noted is the *Elements* in thirteen books (Books 1–6 on plane geometry, 7–9 on the theory of numbers, 10 on irrationals, 11–13 on solid geometry). This great textbook immediately took the place of the works of earlier writers who outlined mathematics. For it put together in better order the work of these geometers and included many propositions which Euclid himself discovered. When Ptolemy asked if geometry could not be made easier, Euclid replied that there was "no royal road."

Archimedes was perhaps the most famous of ancient mathematicians. Born at Syracuse, he studied for some time in Alexandria, where he met Eratosthenes and Conon of Samos, and then returned to his native city; there he composed most of his mathematical works. His mechanical inventions were also numerous and important; they included a pump known as the waterscrew of Archimedes, and the compound pulley. The tradition is that he himself placed little value on these works of applied mathematics, though there is no way to be sure of Archimedes's attitude toward his inventions. He was a friend of Hieron, king of Syracuse, for whom he built various engines of war. Many years later, these weapons were used in the defense of Syracuse against Marcellus, and they changed the siege into a blockade. When Syracuse was finally conquered in 212 B.C., Archimedes was killed by the Roman soldiers. At the time, his mind was earnestly fixed upon a mathematical problem.

Apollonius of Perga, commonly called the "great geometer," was educated at Alexandria under the successors of Euclid during the reign of Ptolemy III. He is known especially for his work on conic sections, the branch of geometry which deals with the ellipse, parabola, and hyperbola.

Archimedes contemplating a problem. This six-teenth-century portrait of the mathematician and inventor was also drawn by André Thevet. It is taken from a large bronze medal discovered in excellent condition beneath a town in Sicily founded by the Romans; the picture therefore may be quite lifelike. Syracuse, Archimedes's birthplace, was the leading city of ancient Sicily. On the table in the picture is what appears to be a representation of the siege of Syracuse and a model of one of the inventor's remarkable war machines.

Conon of Samos pursued astronomy and mathematics in Alexandria in the third century B.C. He named a famous cluster of stars. According to the story, Berenice, wife of Ptolemy III, cut off her amber-colored tresses and placed them in the temple of Venus at Zephyrium, as she had vowed to do if the king returned victorious from the war against Babylon. Soon the hair vanished from the temple, and to console the royal couple for the loss, the astronomer named the star cluster Coma Berenices, or Berenice's Hair. Conon was also well known for his researches into eclipses of the sun. He was a close friend of Archimedes, who praised his mathematical work highly and regretted his early death.

Heron, another able mathematician, was a native of Alexandria who probably lived several centuries after Archimedes. He was the inventor of slot machines, "Heron's fountain," a fire engine, a water organ and numerous devices operating by means of water, steam, or compressed air. And Diophantus of Alexandria, who lived in the third century A.D., is usually given credit for introducing algebraic methods into mathematics.

7 ✫ A Second Alexandrian School

After Alexandria came under Roman control in 30 B.C., the influence of the Alexandrian school was spread over the whole known world. However, with the growth of the empire, men of letters started to gather at Rome rather than at Alexandria. At the same time, Alexandria developed a new movement, which was not at all in the old direction. The character of this second Alexandrian school was determined largely by Oriental mysticism and contained Jewish and, later, Christian elements. It resulted in the speculative philosophy of the Neoplatonists and the religious philosophy of the Gnostics and early Church Fathers.

The Neoplatonists modified Plato's theory of ideas into a doctrine of mystical emanation of the material world from a spiritual principle. Neoplatonism began at Alexandria in the early third century A.D. with the oral instruction of Ammonius, who was nicknamed Saccas (sack bearer) from his early occupation as a porter. His teaching was soon overshadowed by that of his pupil Plotinus, the chief exponent of this philosophy. Plotinus was born in Egypt in A.D. 205 and taught during the latter part of his life at Rome, where he became the center of an influential circle of intellectuals. The Gnostics set forth doctrines which combined Christian theology with Oriental religions and with Neoplatonism.

55

Gravestone with painted decoration. This is one of the most artistic of the painted steles (slabs used as gravestones) found at Alexandria. The painting in the niche represents three figures on a rose-colored ground. A rearing chestnut horse is trying to break away from a young man with a conical cap covering his red hair. The man is wearing a yellowish belted chiton, the garment commonly worn next to the skin by both sexes in classical times, and perhaps a sword. Behind is a red-haired boy dressed in the same manner as the man; he is looking at his master or father. The horse's head is thrown upward, his ears are back, and his eyes are full of fire. Found in the necropolis of Hadra, the gravestone is made of marble.

PART TWO *** THE SUN

IS CENTERED

8 ⁂ *The City Grows*

Returning to the golden days of Alexandrian culture and to the development of astronomy, the focus of this story, one of the greatest astronomers that the world has ever known is encountered, Aristarchus of Samos. He must have been active about the year 281 B.C., since Claudius Ptolemy reported that Aristarchus observed the summer solstice of that year. This observation of the sun's position on June 22, when it was at its most northern point in the heavens, was probably made in Alexandria. What was the city like when Aristarchus of Samos worked there more than 2,000 years ago?

Founded on the site of the townlet Rhacotis, a resort of fishermen and pirates on the Egyptian coast, Alexandria was intended to give Alexander a secure naval base for his operations in Persia and to serve as a link between Macedonia and the rich Nile Valley. The coast was low and sandy, with Pharos island lying off it. Behind Rhacotis were several native villages and Lake Mariotis. So Alexandria lay on a narrow strip of land between the Mediterranean and the lake.

Alexander occupied Pharos and had a walled city marked out by Dinocrates of Rhodes. Dinocrates was perhaps the most eminent architect of his time. He built an enormous funeral pyre for burning the body of Hephaestion, a Macedonian officer who had been a close friend of Alexander. He conceived the idea of carving one of

59

Paintings of Hellenistic buildings. Belonging to a wall decoration, these paintings from the ancient city of Pompeii portray a monumental entrance to a sacred grove *(opposite)* and perhaps an entrance to a royal park *(above)*. Certain peculiarities of the architecture suggest that the buildings do not belong to Roman villas but are constructions of the Hellenistic period, possibly in Ptolemaic Egypt.

the peaks of Mount Athos into the shape of a huge statue of Alexander. He was still alive during the reign of Ptolemy II, and Pliny tells in his *Natural History* that the architect had planned an unusual temple in memory of Arsinoë, the king's wife: the roof of the temple was to be furnished with lodestones so that the queen's statue would appear to be suspended in air.

Not long after the construction of the city was begun, Alexander left Egypt, never to return. His Collector of Revenues for the province, Cleomenes of Naucratis, continued the creation of the city. However, Cleomenes squandered Egypt's wealth in selfish schemes and was finally executed by the first Ptolemy, who brought serious charges against him. Alexandria grew in less than a century to be larger than the celebrated city of Carthage in North Africa, and for the following few hundred years it ranked second only to Rome. Many scholars think that it continued to grow until it surpassed Rome in size. It was a center of Hellenism as well as of Judaism, or Jewish culture; in respect to population, it was the largest Jewish city in the world.

Alexandria was at first divided into three regions: the Jewish quarter; Rhacotis, occupied primarily by Egyptians; and Brucheum, the royal or Greek quarter. Brucheum was the most majestic part of the city. In Roman times Alexandria was enlarged by the addition of an official quarter, making four regions in all.

The royal buildings of the Brucheum were richly adorned inside and out. Some of their walls were built of slabs of colored stones, especially alabaster; others were built of brick covered with slabs of this kind. The rooms contained fine wooden furniture bedecked with figures and ornaments made of ivory and costly metals and covered with beautiful rugs. The floors were decorated with mosaics. No less sumptuous and stately were the tall houses of the aristocracy. And in the royal parks and gardens stood lovely monumental fountains. One of these was built of Hymettian and Parian marble and Egyptian syenite. Dedicated to Ptolemy II and Arsinoë, it was decorated with the royal couple's portrait.

Laid out on the gridiron plan, the city contained fine parallel streets. In its center, two great avenues met and crossed. These were about 200 feet wide, paved with squares of granite and lined with marble colonnades. The Canopic Way ran almost due east and west. Westward it ended at the Necropolis Gate; beyond was the City of the Dead, with gardens, graves, and establishments for the embalming of corpses. Eastward it extended to the Canopic Gate. The street of the Soma started from the Sun Gate by the lake harbor and ran northward to the Moon Gate by the sea. It intersected the Canopic Way in the center of the city, close to the Soma, Alexander's stately tomb.

The residents of the city constituted an unusual society: the king and his court; the army; the high officials; the magistrates and priests of the city; the members of the City Council; the scholars, scientists, poets, writers, and philosophers of the museum and library; young men and schoolboys and girls; Greek and native priests; rich businessmen, modest shopkeepers, artisans, peddlers, longshoremen, sailors, and slaves.

Numerous languages were spoken. Various dialects of Greek were most common. But in the native quarter Egyptian was the language of the inhabitants, whereas in the Jewish part of Alexandria Hebrew and Aramaic were still the predominant tongues. On the streets and in the harbors, other Semitic languages, and perhaps even Indian dialects, might be heard.

At first Alexandria included little more than the island of Pharos. This was connected with the mainland by a mole, or causeway, called the Heptastadion—about one mile long and 600 feet wide. At the point where this causeway joined the land rose the

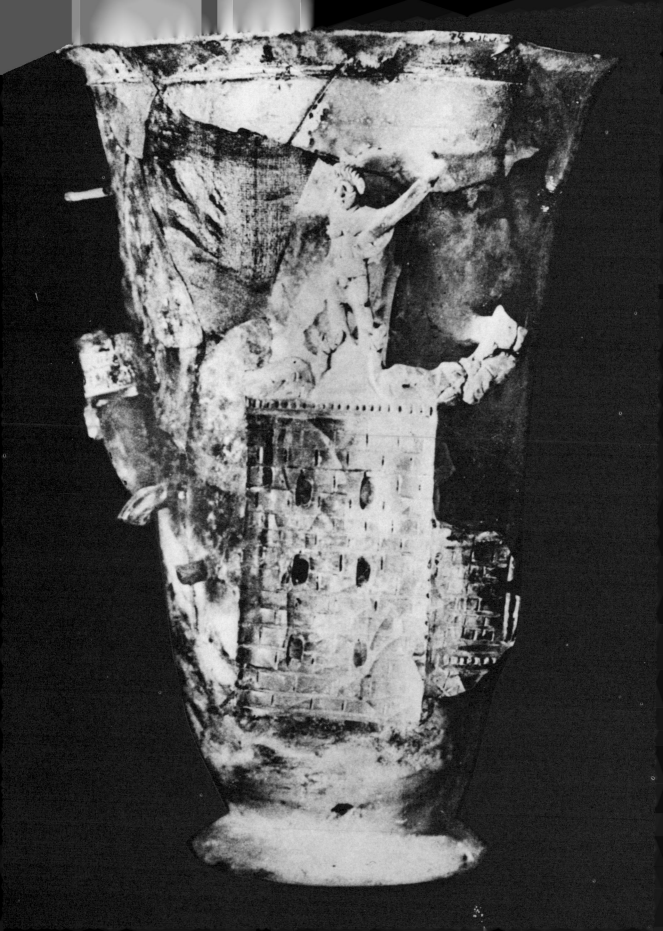

The Pharos depicted on a glass vase. Unearthed at Begram in Afghanistan, the vase is remarkable for the way its figures are executed—they are almost three-dimensional sculpture. Dating from the third to the fourth century A.D., it has been reassembled out of a number of fragments and shows on its face a massive structure surmounted by a naked man. The tails of two sea monsters can be distinguished beneath the man; the higher parts of their bodies are hidden but enough of the creatures remain to show that they had human torsos. The statue at the summit and the figures of the tritons beneath him form a characteristic decoration of the most ancient and famous of lighthouses. The other side of the vase depicts the open sea. The principal categories of ships of the high sea are represented by a war galley with two rows of oars, a merchant ship with a sail rendered with great care, and a modest fishing boat with a decorated hull. The composition of the tower and the ships brings to mind the views of ports encountered in Roman art; these generally show harbor buildings in the background and a number of ships in the foreground (see pages 48-49, 114).

Moon Gate. The Heptastadion and the mainland quarters are believed to have been primarily Ptolemaic work.

The two harbors lying east and west of the causeway were the Great Harbor and Eunostes, or "the haven of happy return." The mainland, the Heptastadion, and the eastern end of Pharos protected the Great Harbor on three sides. On the open (eastward) side, the Great Harbor was protected by a pier built out from the mainland and by a string of reefs which made the harbor difficult to enter.

It was for the purpose of guiding sailors past the reefs to the safety of the harbor that the famous lighthouse of Pharos was planned by the architect Sostratus of Cnidos. One of the seven wonders of the ancient world, the tower, erected upon a heavy stone platform, was made of white marble and rose in decreasing stages for 400 feet (nearly one-third the height of the Empire State Building). The lowest section was square, the middle section octagonal, and the uppermost section cylindrical in shape. The light, probably a fire that was kept burning at night on the top platform, could be seen 300 miles out to sea. The second Ptolemy built the tower, which became the model of all lighthouses constructed thereafter.

65

Serapis and the Pharos lighthouse. The bronze coin represented was minted in Alexandria during the reign of the Roman emperor Antoninus Pius. The lighthouse, on the reverse, appears as a tower larger at the base than the summit. It is pierced by two vertical rows of openings. At the top, there is an allegorical figure standing on a pedestal. To the left and right of the figure, tritons blowing trumpets can be distinguished. Alexandrian money representing the Pharos shows a diversity of forms; it is obvious that the coin makers contented themselves with giving a general impression of the lighthouse and were not preoccupied with details. On the other hand, the glassblower who made the vase found at Begram (see page 64) was able to show details necessarily invisible on small coins, in particular in the structure of the walls and in the form of the windows. Some scholars have believed that the lighthouse had an uppermost story that was cylindrical in shape. Others have believed that this top story was only the base of the statue; this is perfectly clear in the representations on money and the glass vase. The statue is believed to be Poseidon, Zeus, or one of the Ptolemys. The figure on the vase is much larger than the small images on coins, and from its face it seems most probable that the statue portrays Ptolemy I (compare the face on the vase with the marble mask of Ptolemy I on page 33). Made of bronze, the statue is believed to have had a height of 23 feet.

9 ✺ The Sun and Moon Are Measured

Aristarchus wrote on the sizes and distances of the sun and moon under the rule of Ptolemy II. He lived at the same time as Archimedes, though he was older than that mathematician who, according to Pappus of Alexandria in his *Collection,* boasted of being able to move a great weight by a small force: "Give me a place to stand and I will move the earth."

Aristarchus was born on the island of Samos in the Aegean Sea, where the Greek philosopher Pythagoras was born several centuries earlier. He studied at Alexandria under Strato of Lampsacus, the philosopher who was called to Alexandria by the first Ptolemy and acted as teacher to the second Ptolemy. Strato was surnamed Physicus (the physicist) because he devoted himself to the study of science. It was largely due to his influence that the museum became primarily an institute of scientific research.

Aristarchus was known as "the mathematician," probably in order to tell him apart from others with the same name, though *On the Sizes and Distances of the Sun and Moon,* the only work of his that still exists, is the achievement of a very able mathematician indeed.

In the mechanical line, Aristarchus is given credit for having invented an improved sundial called the skaphē. It consisted of a hemispherical bowl with a needle erected vertically in the middle to

67

Banqueting tent of Ptolemy II. Based on a description by Callixenus, a Rhodian Greek who flourished about 155 B.C., this reconstruction is by the Austrian archaeologist and classical scholar Franz Studniczka (1860–1929). Both the banqueting tent and the floating villa of Ptolemy IV (see page 82) were gorgeously decorated with colored and ornamental rugs. From the picture, it is easy to imagine a grand dinner in honor of foreigners at the court of this king. Aristarchus was a contemporary of Ptolemy II.

cast shadows. This device made it possible to read off the direction and the height of the sun by means of lines marked on the surface of the hemisphere.

The first serious attempt to find the sizes and distances of the sun and the moon by observation is found in Aristarchus's only surviving book. Aristarchus observed the angular distance between the sun and the moon at the time of half-moon (that is, when the angle at the moon in the earth-moon-sun triangle is a right angle). He found this angular distance equal to 87°. From this he deduced the result that the distance of the sun was between eighteen and twenty times as great as the distance of the moon. Actually it is nearly 400 times as great. Though theoretically correct, the method is not practical, as the moment when the moon is half illuminated cannot be determined accurately.

Inexact though it was, this calculation drew attention to the difference in the distance of the two celestial bodies. It also showed that the sun and the moon must be of very different sizes. They look the same size in the sky; this can be proved to be true at an eclipse when the sun is hidden by the moon. Consequently, their actual sizes must be proportional to their distances—a ratio which Aristarchus had grasped.

The remaining problem was to discover the actual sizes of the sun and moon. These sizes could be learned from the size of the shadow which the earth casts over the moon at an eclipse of the moon. As the sun is so far away, the earth's shadow must be almost equal to the earth which casts it. Aristarchus figured that the diameter of the shadow was about seven times the diameter of the moon. The actual figure is about four. And although his estimates were not accurate, they showed that the sun must be much larger than the earth.

10 ✵ *"The Ancient Copernicus"*

Eighteen centuries before the Polish astronomer Nicholas Copernicus declared the sun to be the center of the planetary system, Aristarchus originated the heliocentric (sun-centered) hypothesis in astronomy. He therefore deserves his title of "the ancient Copernicus."

One of the reasons that led him to place the sun as the center of the universe was probably his estimation of the relative sizes of the sun and earth. In *On the Sizes and Distances of the Sun and Moon,* Aristarchus finds the ratio of the diameter of the sun to the diameter of the earth to be greater than 19:3 but less than 43:6. This makes the volume of the sun over 300 times the volume of the earth. Although the principles of dynamics were at that time unknown, it might have seemed ridiculous even in the third century B.C. to make the body which was much larger move in a circle around the smaller.

What were the beginnings of the sun-centered scheme? Philolaus of Croton (fifth century B.C.), a Pythagorean philosopher and contemporary of the great Greek thinker Socrates, had already suggested dethroning the earth from its supposed central position. He made it revolve with the other planets and the moon and sun around a central fire which he called "the hearth of the universe."

70

The fourth-century philosopher Heraclides Ponticus, of Heraclea on the Pontus (Black Sea), had taught that Mercury and Venus revolved around the sun, while at the same time the sun and the other planets revolved around the earth. (This scholar, a pupil of the philosopher Plato, is also believed to have been the first to state positively that the apparent rotation of the heavens is caused, not by circling of stars around the earth, but by the turning of the earth upon its own axis.)

Why not, Aristarchus may have thought, put together the two suggestions of Philolaus and Heraclides and imagine that all the planets move about the sun?

If the earth moved in this way, one would expect its motion to cause the fixed stars perpetually to change their directions as seen by an observer on the earth, and Aristarchus may have looked confidently for the changing appearance of the sky. Yet no such change was observed, and he may have realized that this could mean only one thing: the stars are so very far away that the earth's revolution around the sun makes no change that can be seen in their apparent positions.

Archimedes says that Aristarchus published his views in a book, but since this book is no longer in existence, it is necessary to rely on the statements of other writers. In *The Sand Reckoner* (in which Archimedes works out the number of grains of sand which, on certain assumptions, the universe would contain), he gives the following account of Aristarchus's ideas about the universe: "His hypotheses are that the fixed stars and the sun are stationary, that the earth is borne in a circular orbit about the sun, which lies in the middle of its orbit. . . ." The passage goes on to state that Aristarchus held that if the earth is supposed to move around the sun in a large orbit, the distances of the fixed stars must be enormously great as compared with that of the sun.

71

THE SUN IS CENTERED

In *On the Face in the Disc of the Moon* by the Greek biographer and philosopher Plutarch, who wrote about A.D. 100, there is another reference to Aristarchus's hypothesis, which says that Aristarchus ". . . supposed that the heavens stood still and the earth moves in an oblique circle at the same time as it turns round its axis." So the plan of Aristarchus included the rotation of the earth about its own axis.

Terracotta statuette of Isis and Serapis. Isis was the wife, first of Osiris, and later of his successor Serapis. Originally the Egyptian goddess of motherhood and fertility, she also become a goddess of navigation, traversing the sky in a boat. In Ptolemaic times her temple in Alexandria was on the hill on which the Serapeum stood, and she was identified with Arsinoë II and with later queens of the same dynasty.

11 ✩ *The Failure of an Idea*

Aristarchus had arrived at an accurate understanding of the sun's system. He had reached true ideas of the relatively small size of the earth, its apparent unimportance in relation to the far bigger sun, and the insignificance of both of these objects in the immensity of space. From this good beginning, it would be expected that the story of astronomy would be one of swift progress along scientific lines. But this was not the case. Only one man, Seleucus of Seleuceia on the Tigris, supported Aristarchus's heliocentric hypothesis. This was about 150 B.C. Apart from this Babylonian's support in antiquity, the scheme never attracted much attention until the time of Copernicus and later of the Italian astronomer Galileo Galilei.

Aristarchus's bold and imaginative idea was undoubtedly too much opposed to the general views of the various schools of thought to obtain a serious hearing. According to Plutarch, Cleanthes, head of the Stoic school of philosophy and a contemporary of the astronomer, thought ". . . that Aristarchus of Samos ought to be accused of impiety for moving the hearth of the world. . . ." (*On the Face in the Disc of the Moon*). At the time, astrology, the investigations of the aspects of the planets and their imagined influence upon the destinies of men, was very popular. To many intelligent people, it was a respectable science, and its doctrines and

73

findings were based upon a geocentric (earth-centered) hypothesis of astronomy.

The main reason the heliocentric idea failed to make a strong impact, however, was the rapid rise of practical astronomy. When more and more of the apparently complex motions of the planets began to be noticed, the hopelessness of trying to account for them by the beautifully simple idea of Aristarchus must have given the fatal blow to his purely speculative system.

Many centuries later Copernicus was obviously disappointed to learn that Aristarchus had preceded him in proposing a sun-centered universe. In the handwritten copy of *On the Revolutions of the Celestial Orbs*, the Polish astronomer cut out a statement admitting his knowledge of Aristarchus's hypothesis; that is, he intentionally withheld the statement from publication. Elsewhere he tells of finding in Plutarch the views of such men as Philolaus and Heraclides, but he does not mention the reference to Aristarchus's hypothesis that also appears. Copernicus is also believed to have been acquainted with Archimedes's *The Sand Reckoner*, which, of course, contains the best account of Aristarchus's inspired guess.

Aristarchus died in 230 B.C. Some five years earlier Ptolemy III had appointed as head of the library of the great museum in Alexandria the versatile scholar Eratosthenes, one of whose accomplishments was to measure the earth.

PART THREE ✷✷✷ THE EARTH
IS MEASURED

12 ✲ A Celebrated Librarian Is Born

The Delphic oracle was the principal center of the worship of Apollo, which was held in exalted honor by all Greeks in remote antiquity. Its site was supposed to be the center of the earth, marked by the sacred navel-stone. Here, it was believed, the god Apollo, through his priestess, called the Pythia, foretold the future. According to a late tradition, the Pythia seated herself upon a golden tripod over a chasm from which intoxicating vapors came. Inspired by these, she uttered words which were then arranged by prophets especially trained for the purpose. Many of the replies of the oracle were capable of being understood in more senses than one. This gave rise to the phrase "delphic response."

One of the replies of the oracle brought about the founding of Cyrene, Eratosthenes's birthplace. Inhabitants of Crete and Thera were encouraged to settle in the north of Africa about 630 B.C. In the fourth century B.C., Cyrene, lying between Alexandria and Carthage, was renowned for its philosophers and mathematicians. Alexander made it an ally, and Ptolemy joined it to his Egyptian kingdom.

Eratosthenes, son of Aglaös, was born in this city in 276 B.C. At some period during his early manhood, he went to Athens for

the equivalent of a university education today. There he studied philosophy under the Stoic philosopher Ariston of Chios and the Skeptic Arcesilaus of the Academy (this philosopher is said to have died in his seventy-sixth year from a fit of drunkenness).

When he was about thirty, Eratosthenes was invited to Alexandria by Ptolemy III. Ptolemy may have been spurred to this action by the grammarian and poet Callimachus, who, it will be recalled, was also born in Cyrene and who had been given a post in the library by Ptolemy II. Callimachus was on close and friendly terms with the courts of Ptolemy II and Ptolemy III and their consorts.

It has already been mentioned that Eratosthenes was appointed head librarian about 235 B.C. He became tutor to Ptolemy III's son at some point during his stay in Alexandria, and he continued in favor with the royal court until he died.

Libation vessel portraying Berenice II. This piece of pottery, with greenish-blue glaze and traces of gold and painting, was found at Bengasi (ancient Berenice, a coastal city of Libya). The relief decoration shows Queen Berenice II, wife of Ptolemy III, wearing a royal diadem and holding in her left hand a cornucopia full of fruit and ears of corn. She is pouring a libation over a rectangular horned altar decked with a garland. Behind Berenice there is a betyl (a conical sacred stone symbolic of a deity) on a cylindrical base. Pottery of this type seems to have been sold in Egypt and the Ptolemaic dominions in shrines and temples devoted to the royal cult and used for libations to the deified rulers, then taken home as souvenirs. Since all the known pottery of this kind apparently belongs to the third century B.C., and only Arsinoë II, Berenice II, and Ptolemy IV are mentioned in the inscriptions on the pottery, the fashion must not have lasted long.

13 ✲ "A Sanitorium for the Mind"

If you had visited the magnificent library at Alexandria, you would have seen on the entranceway the inscription: "A sanitorium for the mind." So writes Diodorus Siculus of Agyrium, in Sicily, the celebrated historian of the late first century B.C., whose work in 40 books was entitled *The Historical Library*. A sanitorium is a place to which one goes to be cured, and the reference is plainly to the library as a whole.

There were two libraries at Alexandria, and they were the most important in the ancient world. The larger one, with the lettering on the entranceway just described, was in the Brucheum quarter, attached to the museum. The smaller was placed in the Serapeum, the most famous of all Alexandrian temples. It was created in the middle of the third century B.C.; by then the main library was so large that a "branch" was found to be necessary. These libraries included the literature of Greece, Rome, India, and Egypt. According to information drawn from Callimachus and Eratosthenes, there were 42,800 rolls of papyrus or volumes in the Serapeum and 490,000 in the Brucheum. As the ancient volume usually contained a much smaller amount of matter than a modern book, less than full credit must be given to these figures for the purposes of comparison with libraries of the present time. The *Iliad* of Homer, as pointed out before, consisted of twenty-four "books."

By way of comparison, the Library of Congress, the national library of the United States and perhaps the greatest library in the modern world, had a total of 25 million books, pamphlets, manuscripts, recordings, music, photographic prints, and so on, by mid-twentieth century. Of this number about 9 million were volumes and pamphlets. The New York Public Library, an endowed institution and another of the great libraries of the world, had at mid-twentieth century over fifty branches supported by city funds and almost 5 million volumes.

There is considerable uncertainty about the order in which the early librarians at Alexandria followed one another. A papyrus fragment (*P. Oxyrhynchus* 1241) lists them in the order given here, but there are several errors and chronological difficulties in this series of names. The list begins with Zenodotus, who became the first head of the library at Alexandria about 284 B.C. He was a pupil of Philetas, an Alexandrian poet and grammarian who suffered from bad health and was abnormally thin. Philetas had other famous pupils, including Ptolemy II.

Apollonius of Rhodes, who was a pupil of Callimachus, followed Zenodotus as librarian about 260. Callimachus had many critics, among whom was finally included his former pupil Apollonius. Those who opposed him said that he did not have enough ability to compose one continuous poem in many thousands of verses (that is, an epic in the customary manner). Callimachus replied that such writings were out of date, and he limited himself to the composition of short poems complete in themselves or loosely connected in a larger work. In his later years, Apollonius's epic, the *Argonautica*, in four books, marked a rebellion in principle against

Callimachus and provoked a famous quarrel. The quarrel may also have been the result of lack of harmony at the library, where it is fair to suppose that Callimachus's official position was inferior to that of Apollonius. At any rate, Callimachus was victorious, and Apollonius retired to Rhodes where he stayed till his death.

The next head of the library was Eratosthenes, who had also been a pupil of Callimachus. Aristophanes of Byzantium—a pupil at Alexandria of Zenodotus, Callimachus, and Eratosthenes—was Eratosthenes's successor; this was about 194 B.C. Aristarchus of Samothrace, a pupil of Aristophanes, probably preceded another Apollonius as head of the library about 153 B.C.

Ptolemy IV's barge of state. The most famous ships of the ancient world were those of Ptolemy IV, contemporary with Eratosthenes. The king had this "cabin-carrier" constructed for his trips on the Nile in the late third century B.C. The ship had a length of 300 feet, a breadth at the widest part of 45 feet, and a height (including the pavilion) of about 60 feet. It had a double bow, as well as a double stern which extended upward considerably because the waves of the river often rise very high. The hold in the middle of the ship contained saloons for dining parties and other conveniences of living. There were promenades running around the ship on three sides, on an upper and a lower deck. The structure of the lower one had a complete system of roof-supporting columns like those used on the sides of a building. The one on the upper deck was enclosed with walls and windows between the columns. At the stern there was an entrance court, open in front but having a row of columns on the sides; in the part which faced the bow, a main entrance was built. Inside the pavilion, there were cabins provided with couches, dining saloons, sleeping apartments with berths, and a shrine of Aphrodite containing a marble statue of the goddess. Cedar and cypress, ivory and gold, marble, and red copper gilded in the fire—these are the materials with which the rooms of this luxurious barge were made and ornamented. The mast of the ship was 105 feet high; it sustained a sail of fine linen, reinforced by a purple top sail.

The first experiments in bibliography seem to have been made in producing catalogs of the Alexandrian libraries. By order of Ptolemy II, two catalogs were prepared—one of the tragedies, the other of the comedies. Callimachus, who was employed as cataloger, produced a very large catalog of all the principal works. Called Tables of the Outstanding Works in the Whole of Greek Culture and of Their Authors, it filled 120 volumes. The books were divided into eight classes: (1) dramatists, (2) epic and lyric poets, (3) legislators, (4) philosophers, (5) historians, (6) orators, (7) rhetoricians, (8) miscellaneous writers. The scientific books may have been placed in either the fourth or the eighth class.

83

14 ✺ The Tilting of the Earth's Axis

Eratosthenes was one of the most eminent men of learning of his time. He produced works on geography, mathematics, philosophy, grammar, and, as already noted, literary criticism and chronology. Indeed, he ranked so high as a mathematician in the estimation of Archimedes that Archimedes dedicated one of his treatises to him and sent him a difficult problem for communication to the mathematicians of Alexandria. And he wrote poetry. The *Hermes*, a short epic, described the birth of the god, his noteworthy youthful deeds, and his ascent to the heavens. The *Erigone*, an elegy, was a star legend telling the story of Icarius, his daughter Erigone, and her dog. In this version they were all conveyed to the heavens where they became constellations. The short epic *Anterinys*, with the alternative title *Hesiod*, is believed to have dealt with the death of the Greek poet Hesiod and the punishment of his murderers. Only a few fragments of Eratosthenes's poetry are still in existence, and it is impossible to judge from this his merit as a poet.

Polymathy—wide and varied learning—was greatly respected by the Alexandrians, and Eratosthenes was called "All-rounder" by his admirers. He was also called "Beta" (the second letter of the Greek alphabet, corresponding to English *b*). He may have been given this latter name because, working in so many fields of knowledge, he was second-best in all fields but first in none.

Eratosthenes is given credit for having invented the armillary sphere. This is a skeleton sphere composed of rings. It is designed to show the positions of important circles of the celestial sphere—the spherical surface on which the heavenly bodies appear to lie. Eratosthenes had *armillae* (armillary spheres) installed in the portico of the museum, and he made various measurements with these astronomical machines. One of these was the obliquity of the ecliptic—that is, the tilting of the earth's axis of rotation which causes the seasons. He arrived at the value of 11/166 of a complete circle, or 23°51′—a remarkably accurate measurement, since the true value at the time was approximately 23°46′.

OVERLEAF. *The Letronne Papyrus.* The oldest illustrated Greek papyrus known today is an astronomical one. Dating from the second century B.C., it contains instructions about the celestial spheres that are based on propositions by Eudoxus, a Greek scholar of Cnidus, who lived in the fourth century B.C. Crudely drawn diagrams of the constellations and the zodiac are distributed throughout the columns of text. Some of the illustrations are more than mere line drawings: one shows a figure of the god Osiris in a disk representing the constellation Orion, another a scarab as the symbol of the sun in the center of a diagram of the zodiac. These signs demonstrate the influence of Egyptian astronomical texts; the Greek illustrator must have known these texts and copied the images from them. Such iconographic details strongly support the theory that Alexandria played a significant role in Hellenistic book illustration.

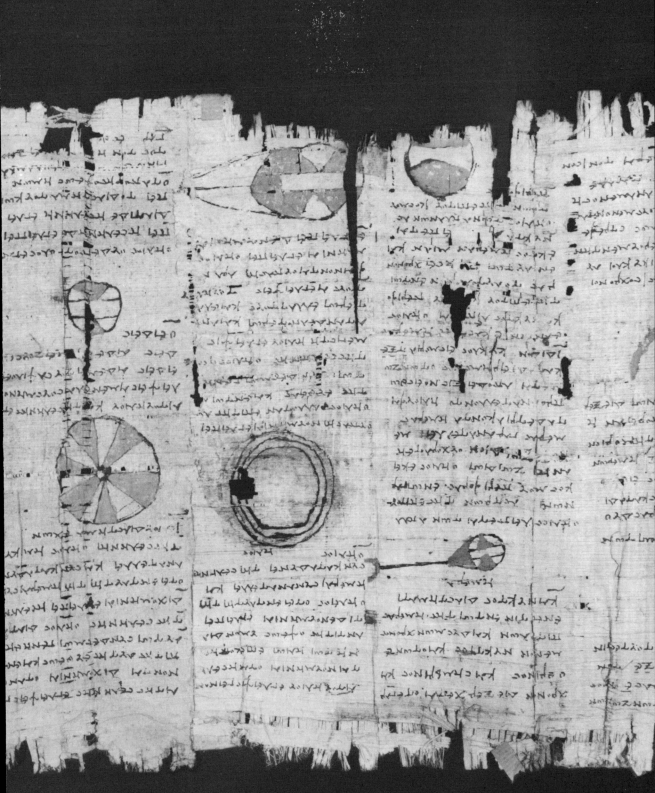

15 ✦ *The Well of Light*

Eratosthenes's most famous achievement was his measurement, made one midsummer day, of the circumference of the earth. This may not have been the greatest achievement of Alexandrian science, but it was certainly the most thrilling.

The method by which Eratosthenes determined the distance around the earth is notable both for its simplicity and for the precision of the result obtained. It was based on the observation that a well at Syene (modern Aswan), a frontier outpost under the Ptolemies, was lighted to its remotest depth at noon on the summer solstice—the time of the year when the sun is at its greatest angular distance northward from the celestial equator. From this it followed that on that day the sun passed exactly through the local zenith, the point in the sky directly above the observer at Syene. The well, incidentally, was dug especially for the purpose and still exists.

At the same instant, in Alexandria, Eratosthenes made a measurement of the shadow cast by a needle fixed in the center of a hemispherical bowl. This instrument was the skaphē invented by Aristarchus. Eratosthenes estimated that the shadow amounted to 1/25 of the hemisphere and thus 1/50 of the whole circle. The sun, therefore, was 1/50 of a complete circle (or 7°12′) below the zenith at Alexandria.

The well of Eratosthenes. The well traditionally ascribed to Eratosthenes is on the Island of Elephantine at Aswan on the Nile, in upper Egypt. It is approximately 25 feet deep, with spiral steps leading down to the water. The upper part of the stonework is modern, but the projecting stones below the modern wall were used formerly to hold up a roof, and everything below them is old. The photograph was taken in 1914.

Eratosthenes believed that the two places were on the same meridian, or north-and-south line passing through the poles, and concluded that the earth's surface at Syene made an angle equal to 1/50 of a complete circle with the earth's surface at Alexandria. From this it followed that the circumference of the earth must be fifty times the distance from Syene to Alexandria.

Professional "steppers" or "pacers" found this distance to be 5,000 stadia, which made the earth's circumference 250,000 stadia.

THE EARTH IS MEASURED

It appears fairly certain that Eratosthenes later changed his estimate to 252,000 stadia.

The question is, what was the length of the stadium used by Eratosthenes? The precise length is unknown, but it is likely that it was about 517 feet. If this length is assumed, the distance around the earth comes out at approximately 24,650 miles—a figure that did much to encourage the growing belief that the earth is very large. The true value is 24,875 miles. Although Eratosthenes seems to have made his measurements only in round numbers, so that to some degree the extraordinary accuracy of his final result must be attributed to genuine good fortune, this was still a remarkable achievement for the time.

In his old age, Eratosthenes lost his eyesight, and, growing weary of life, he is said to have committed suicide by voluntary starvation. The scholar was then in his early eighties. The year was 194 B.C.

Hairpin depicting the goddess Aphrodite arranging her hair. Bronze models of toilet articles, chiefly hairpins with tops in the form of figures or other ornaments, were found at Galjûb near Cairo. Of the second century B.C., they served as samples in the shop of a gold- and silversmith and were reproduced in precious metals or ordered by the artisan's customers. A business of this kind was conducted in the workshop of a certain Mystharion in Alexandria in early Roman times. The city of the stargazers was an important center of metallurgy.

16 ✦ *Island of the Sun God*

During the last century and a half before the Christian era, as was pointed out earlier, Rhodes was to a great extent a rival of Alexandria as a center of literary and intellectual life.

The most easterly island of the Aegean sea, with an area of about 545 square miles, Rhodes was believed sacred to Helios, the sun god. According to legend, the god himself chose the island, which had not then risen above the surface of the sea. His children by the nymph Rhodos were directed by Helios to offer sacrifice to Athena the day she was born, but they forgot to bring fire. In this way, the custom of fireless sacrifice originated—each year a four-horse team was thrown into the sea to celebrate the festival called Halieia. The sons of Helios were Ialysus, Lindus, and Camirus, after whom the leading cities of the green and mountainous island were named.

The city of Rhodes was founded in 408 B.C. as the capital of the federated Rhodian city-states. It was laid out in an exceptionally fine site according to a scientific plan of the architect Hippodamus of Miletus, who has been described as the originator of city planning. This town soon rose to considerable importance and attracted much Aegean and Levantine commerce which had till then been in the hands of the Athenians. In 332 Rhodes yielded to Alexander, but upon his death the people forced out the Macedonian garrison and from this time on not only maintained their independence but attained great political influence. In 305–304 B.C., there was a famous siege by the Macedonian prince Demetrius Poliorcetes (besieger), son of King Antigonus, who tried without success to force the city into active alliance with his father. Ptolemy I gave assistance to the Rhodians at the time, and this help won him divine honors in Rhodes and the surname of Soter, or savior.

In order to commemorate the defense of Rhodes against Demetrius, a bronze statue of Helios was erected. The cost of making it was defrayed by selling the engines of war left by Demetrius after the siege. About 120 feet high and one of the seven wonders

Head of Demetrius Poliorcetes. On the obverse of a drachma from the mint of Ephesus appears the diademed and horned portrait of the "besieger of cities." Although slightly idealized, it depicts forcibly the individual characteristics of Demetrius, who was proud, bold, and impatient. The horn is the symbol of royal strength and power.

Head of Helios. This tetradrachma was minted in Rhodes, about 304–189 B.C. On Rhodian coins of the fourth century B.C. Helios is shown with rounded face and great locks of hair; the hair is windblown and the crown of rays is merely hinted at by a skillful adaptation of the locks. But in the next century the artists preferred to give emphasis to the rays in a more stylized form. The head shown here may give some idea of the style and aspect of the colossal statue in the harbor that was completed in 281 B.C.

OVERLEAF. *Artist's conception of the Colossus of Rhodes.* Maarten van Heemskerck (1498–1574), famous Dutch religious painter, depicted the Colossus in *The Seven Wonders of the World and the Ruins of the Colosseum.* In the foreground the monument is being constructed. The Latin verse in classical hexameters beneath the picture describes the Colossus: it is made of bronze; within it there is a vast hollow space as in a cavern of stone; it receives divine honors from the Rhodians. A monument of such height and shape was very fragile. Following the earthquake that destroyed it, the fragments are said to have remained on the ground for nearly nine centuries. Then they were sold to a merchant of Emesa (modern Homs), a city of western Syria, who carried them off on 980 camels. (The number of camels varies in different versions of the story from 900 to 30,000.) It is difficult to believe that a bronze mass of this size would be overlooked for so long a time. Compare Maarten van Heemskerck's conception of the Colossus with the coin on this page.

Fialli Fecit   Heemskerck, Inut.

SEPTINOS DECIES CVBITOS ÆQVARE COLOSSVS
DICTVS, PAR TVRRI MOLE SVB NOMINE SOLIS

LIS.

CAVO FACTVS, SAXORVM VASTA CAVERNA
APVD RHODIOS SACROS ACCEPIT HONORES.

4.

of the ancient world, the Colossus took the sculptor Chares, a native of Lindus, twelve years to complete. It stood close to the harbor, but its exact position is unknown. From the sixteenth century A.D. it was generally accepted that the statue had straddled the harbor entrance, with a beacon in its hand and ships passing between its legs. The Colossus was finally destroyed by an earthquake about 224 B.C.

In the days of its greatest power, Rhodes, with its regular streets, well-ordered plan, and numerous public buildings, became famous as a center of the arts. Protogenes of Caunus, who was noted for the extreme care which he gave to every detail of his paintings, resided at Rhodes and embellished the city with his work. His pictures included "Ialysus," "Resting Satyr" (painted during the siege of Rhodes), "Alexander and Pan," and portraits of Aristotle's mother and King Antigonus. It was the home of the poet Apollonius, who, it will be remembered, retired to Rhodes, and of the Stoic philosopher Posidonius. This philosopher of the early first century B.C., born at Apamea on the Orontes, was a man of almost universal curiosity. He taught at Rhodes and wrote works on such subjects as astronomy, history, geography, and moral philosophy. It was the birthplace of Panaetius, the Stoic philosopher of the second century B.C. It had a school of eclectic oratory whose chief representative was Apollonius Molon, a native of Alabanda. This rhetorician of the first century B.C. was the teacher of Cicero and Julius Caesar.

Among the men whose work threw luster on the island of the sun god at this time was a great astronomer. Hipparchus was born too late to see the Colossus of Rhodes standing at the harbor entrance. Nonetheless, its huge fragments, which were not removed until Rhodes was conquered by the Saracens in A.D. 656, must have excited his wonder.

17 ✲ *The Wobbling of the Earth*

Nicaea, best known because the first general council of the Christian church was held there in A.D. 325, was one of the principal cities of ancient Bithynia in northwest Asia Minor. There Hipparchus was born about 190 B.C., a few years before the death of Eratosthenes. Hipparchus made astronomical observations in Rhodes from 146 to 127. Although some scholars have stated that he may have made earlier observations in Alexandria, there is no good evidence that he ever worked there. He died about 120, in a place unknown.

All Hipparchus's original writings are lost, except for a commentary written in his youth on the *Phainomena* of Eudoxus of Cnidus and on the astronomical poem that Aratus of Soli had derived from it. Eudoxus, a brilliant mathematician and astronomer of the early fourth century B.C., described the heavens in the *Phainomena*. Aratus, who flourished in the first half of the third century B.C., turned this prose treatise into verse. The poem achieved immediate fame and many commented on it. In his work, Hipparchus corrected some of the errors of the *Phainomena*—for example, the belief that there was a particular star at the North Pole.

His study of Aratus may have been the beginning of Hipparchus's interest in astronomy, but he soon realized the need of doing something of greater value. He began to make observations of the heavens. As he made more and more of these, his precision increased. The determination of star longitudes and their comparison with earlier longitudes of the same stars led this watcher of the skies to one of his outstanding achievements—the discovery of the precession of the equinoxes.

The precession of the equinoxes is a slow but continual shifting westward of the equinoctial points. The equinoxes are the two points on the celestial sphere where the equator intersects the ecliptic, the apparent path of the sun around the sky during the year. They are called by this name because when the sun in its annual course arrives at either of them, day and night are equal throughout the world. The point where the sun crosses the equator going north is known as the vernal (spring) equinox, and the opposite point, through which the sun passes going south, is the autumnal equinox.

Rock-cut bas-relief of the stern of a Rhodian warship. Still standing in its original position near the ancient stairway leading to the acropolis at Lindus in eastern Rhodes, this beautiful bas-relief was carved, as the inscription on the side of the ship shows, as the base for a bronze statue of a Rhodian naval officer named Hagesandrus. From inscriptions at Lindus, the family of the officer is known. His grandfather was priest of Poseidon Hippios, the Lord of Horses, in 239 B.C.; therefore, the officer's career must be dated at the early second century B.C. and the erection of the statue at about 180 B.C. The dates attributed to the sculptor, who was probably responsible for the base of the statue as well, correspond with this. It is assumed that the honor was conferred on Hagesandrus because of military achievements, which may have been in a successful expedition against the pirates or one against the Rhodians' neighbors and enemies, the Lycians. The drawing of the bas-relief *(below)* shows the details more clearly.

THE EARTH IS MEASURED

In astronomy, the longitude of a celestial body is the distance of its projection upon the ecliptic from the vernal equinox, counted in the direction west to east from 0° to 360°. If the vernal equinox were a fixed point, the longitude of a star would never vary. This, however, is not the case. It has been found that apparently all the stars have changed their places since the first observations were made by the astronomers of antiquity. Change of place here means change of position of the sphere of the stars as a whole to certain fixed coordinates, not change of place of the stars among themselves so as to alter the figures of the constellations.

The explanation given to account for this phenomenon is that the equinoxes have a retrograde motion from east to west, thereby causing the sun to arrive at them sooner than it otherwise would had these points remained stationary. The annual amount of this motion is very small; it is only equal to 50.2″. Since the circle of the ecliptic is divided into 360°, it follows that the time taken up by the equinoctial points in making a complete revolution of the heavens is approximately 26,000 years.

The discovery of precession dates from about 130 B.C. It was detected by Hipparchus by means of a comparison of his own observations with those made by Timocharis, whose observatory was probably part of the museum of Alexandria, at the beginning of the third century. It would have been easier for this extraordinary man to discover the precession if older observations had been available to him, and accurate observations made by Babylonians may have been. Although Hipparchus discovered the precession of the equinoxes and even measured it, he did not understand it; he believed the phenomenon was due to a very slow eastward rotation of the sphere of the fixed stars around the poles of the ecliptic. It was Nicholas Copernicus who, in *On the Revolutions of the Celestial Orbs*

100

(1543), showed that the westward movement of the equinoxes could be explained by a wobbling motion of the earth like that of a spinning top, the axis of which is not quite vertical. And it was Sir Isaac Newton, the English natural philosopher and mathematician, who explained the physical cause of the phenomenon of precession in his *Principia*, first published in 1687. It is due to the gravitational attraction of the sun and moon on the equatorial bulge of the earth.

Hipparchus also compiled a famous star catalog which was published in 127 B.C. It is believed that the catalog consisted of a description of the constellations with some numerical data concerning distances between the stars. The story is that Hipparchus was prompted to compile this catalog by his discovery of a new star that appeared in the heavens. This may be true, although whether the new star was a comet or a real nova (a star which suddenly flares up in the sky and fades away again to its former brightness after a few weeks or months) is unknown. The number of stars in the catalog is also not definitely known; one figure is 850 or less.

Finally, Hipparchus studied the motions of the sun, moon, and planets across the heavens and arrived at results of great accuracy. He gave the length of the lunar month to within less than one second and that of the solar year with an error of only six minutes. In making good measurements of most of the fundamental quantities, he established quantitative astronomy on a reasonably exact basis. He did not construct a theory of the planets but contented himself with showing that existing theories did not satisfy the observations.

In addition to his astronomical activity, upon which his chief claim to fame rests, Hipparchus wrote works on mathematical geography, mathematics, astrology, and weather signs. His *On Chords in a Circle* very likely contained a "table of chords," the Greek equiva-

lent of a table of sines, and initiated the development of trigonomet-
rical methods.

Sound methods, inventive genius, and true scientific proce-
dure characterized Hipparchus's astronomical activity. When he
died, he left behind him a mass of valuable observational material.
But there was no disciple to continue his tradition. Except for a few
observations, the new science of astronomy was put aside for almost
three hundred years. Then Hipparchus's constructive labors attained completion at Alexandria. His observations were employed
by Claudius Ptolemy.

Over the centuries there developed the point of view that
Ptolemy was not really a great man or a highly original man. But it
is now believed that the downgrading of Ptolemy is closely associated with the upgrading of Hipparchus (even in the most casual
discussion of ancient astronomy Hipparchus is called "the greatest
astronomer of antiquity"). The fact of the matter is that practically
nothing is known about Hipparchus. Most of what is known about
him comes from Ptolemy himself, who frequently quotes him, and
in some cases it is impossible to say who the real inventor was. The
downgrading of Ptolemy and the consequent upgrading of Hipparchus may well be an effect of the Copernican Revolution, because
after Copernicus everything geocentric and Ptolemaic came to be
thought of as bad. It would appear, then, that Hipparchus was not
quite so great as earlier scholars have thought and that Ptolemy
should receive some of the credit for originality and achievement
previously given to Hipparchus.

PART FOUR ✦✦✦ THE EARTH
IS RECENTERED

18 ⁂ *A Dynasty Ends*

Although no results of great value were produced in astronomy in the three centuries that elapsed between the time of Hipparchus and that of Claudius Ptolemy, certain dramatic events took place in Alexandria that affected the life of Egypt.

Ptolemy X (or XII), called Ptolemy Alexander II, who had been made joint ruler with and husband of his stepmother, Cleopatra Berenice, murdered his stepmother nineteen days after the wedding. Even though he was the last legitimate male descendant of the dynasty, he was killed by the enraged Alexandrians. His testament, which might not have been authentic, gave Egypt to Rome. This was in 80 B.C., although Alexandria had been under Roman influence for more than a century before.

In 48 B.C. Julius Caesar defeated his rival Pompey at the battle of Pharsalus. Pompey fled to Egypt, where he was put to death by order of the ministers of the teen-aged king Ptolemy XII (or XIV). He got into a boat which the Egyptians sent to bring him to land. But just as the boat reached the shore, and Pompey was stepping on land, he was stabbed in the back. His head was cut off and brought to Caesar, who had pursued him to Egypt. But Caesar turned away from the sight, cried at the death of his rival, and put the treacherous murderers to death.

Head of Pompey the Great. The best-known and best-preserved sculptural portrait of the Roman general who fled to Egypt for protection but was murdered by order of Ptolemy was made into a bust in modern times. It shows Pompey's main features as they are known from coins minted in Sicily and Spain soon after his death and from the numerous references in contemporary and later literary sources: the huge, almost square head, the low forehead with three deep wrinkles, the rebellious curls on the forehead, the small eyes, the short nose and full cheeks, the double chin, and the short massive neck. Some scholars believe this head is a contemporary portrait of Pompey, a work of late Hellenistic art. Others think it is a portrait study made in the time of the Roman emperor Hadrian, based on earlier portraits and on literary sources.

106

Head of a statue of Julius Caesar. The Roman dictator who aided Cleopatra and brought her to Rome is depicted in full military dress. Made by a sculptor of the time of the emperor Trajan, the statue is based on a careful study of the numerous portraits of Caesar then existing; no originals or copies of statues from his own time have survived.

In that same year Caesar was involved in a war against Ptolemy and the Alexandrians. It arose from the determination of the Roman general and statesman that Cleopatra—who had herself carried to Caesar in a carpet or a bedroll—should rule in common with her brother Ptolemy. It ended in the establishment of Cleopatra, who had charmed Caesar, as queen. The young king was defeated by Caesar and drowned in the Nile during flight.

Some years later another Roman general, Marcus Antonius (Mark Antony), entered into a love affair with this exceedingly ambitious and attractive woman. The association of Antony and the queen of Egypt was unpopular at Rome, and Octavian, Julius Caesar's grandnephew who was destined to become the first Roman emperor, declared war upon the foreign queen and ultimately defeated her at Actium in 31 B.C.

Antony committed suicide before Octavian's entry into Alexandria. Cleopatra tried to win the love of Octavian. But when she learned that he had ordered her sent to Rome, to be led as a captive in the procession celebrating his victory, she committed suicide also, probably with poison, but legend says by bringing an asp into contact with her breast. With Cleopatra's death in 30 B.C. the dynasty of the Ptolemies ended, and Alexandria became the capital of a Roman province. Eventually Octavian, who was to receive the title "Augustus" from the Senate, held Egypt as a piece of personal property.

From the time Octavian placed a prefect, or magistrate, from the imperial household over Alexandria, the city appears to have gained anew its old prosperity, for it commanded an important granary of Rome. The importation of grain was essential for the existence of the growing Roman population; approximately 20 million bushels were imported each year. Some of this grain was produced on the emperor's lands in Egypt, but most of it was exacted as tribute from the province.

Head of Cleopatra. Belonging to a set of Ptolemaic coins of Ashkelon, an ancient seaport lying to the northeast of Egypt, this tetradrachma portrays the head of Cleopatra on the obverse. The coin probably dates from 49 B.C. and testifies to the strength of Ptolemaic influence on that free city. The queen's strong and almost forbidding features do not correspond to her popular image, but Plutarch confirms that her beauty was not such as to strike those who saw her. Her irresistible charm lay in her individuality and her powers as a conversationalist.

Head of Mark Antony. This portrait is on the obverse of a Roman gold coin (aureus) of about 40 B.C. Of fine physique, and with a constitution which hardships and excesses alike failed to ruin, Mark Antony was a natural soldier and a great leader of troops. Cleopatra bore him three children.

There was another side to the trade of Alexandria. Expensive rarities from Arabia and India reached Egypt by way of the Red Sea. There is much evidence of the scope of this business intercourse in the first two centuries A.D. The demand for Eastern luxuries was so large that it resulted in a drain of silver from the West to pay for these goods; proof of this is to be found in the large numbers of coins of the early emperors of Rome that have been discovered in the southern part of India.

Although the world of the second century when Claudius Ptolemy lived fell far short of perfection, there was a limited amount (but more than ever before—or since) of law and order and peace among the nations of the earth. This period marked the climax of the golden age of the Roman empire and unquestionably the golden age of Roman science. But the best of Roman science was really the work of such men as the Greco-Egyptian astronomer, geographer, and mathematician Claudius Ptolemy and the Greek physician Galen.

Galen was born in Pergamum about A.D. 129 and, in a spectacular career, rose from gladiators' physician in Asia Minor to court physician in the Rome of the emperor Marcus Aurelius. In his search for knowledge, he visited the medical school at Alexandria. Galen wrote all his life, first philosophical treatises and finally medical books. His research in physiology, based on experiment, was masterly. He proved that the arteries as well as the veins carry blood (the Alexandrian school had taught that the arteries contain air). His research in anatomy was unrivaled in ancient times for its fullness and accuracy.

Galen died in 199; he was the outstanding scientist of the second half of the second century. Ptolemy was the outstanding scientist of the first half.

19 ⁂ *Alexandria from Shipboard*

What was Alexandria like in the days of Claudius Ptolemy? Diodorus Siculus, who, it will be recalled, wrote world history in the first century B.C., estimated that the population of free citizens in 58 B.C. was 300,000. This number could be more than doubled if the slave population were added.

Strabo, historian and geographer of Amasya, Pontus, was in Egypt about 25–19 B.C. collecting geographical material. In his time, the principal buildings of Alexandria, named one by one as they could be seen from a ship entering the Great Harbor, included the following:

The royal palaces filled the northeast angle of the town and occupied the promontory of Lochias, a high point of land reaching into the sea which shut in the Great Harbor on the east.

The great theater was close to the palace system.

A little distance from the theater and in front of it was the Poseidium or temple of the sea god.

The Timonium was next. This was a hermitage built in the Western Harbor by Mark Antony in imitation of the famous misanthrope, or man-hater, Timon of Athens, who lived in the late fifth century B.C.

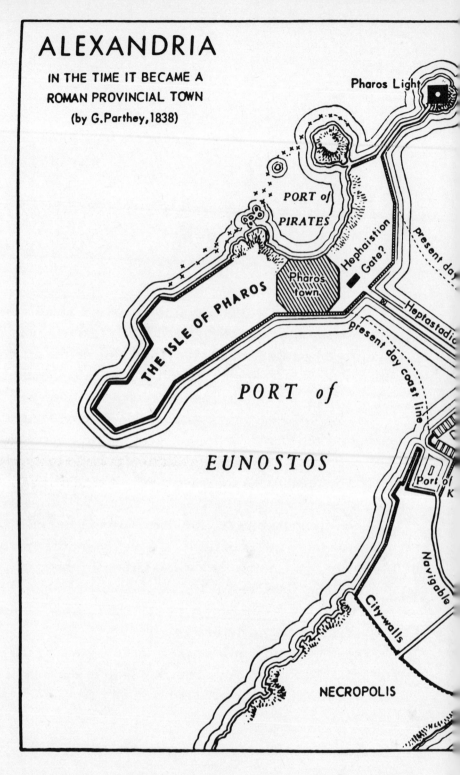

ALEXANDRIA

IN THE TIME IT BECAME A
ROMAN PROVINCIAL TOWN
(by G. Parthey, 1838)

Pharos Light

PORT of
PIRATES

Hephaistion
Gate?

present day

Pharos
town

Heptastadion

THE ISLE OF PHAROS

present day coast line

PORT of

EUNOSTOS

Port of
K

Navigable

City-walls

NECROPOLIS

Map of ancient Alexandria. Delineating the city in the time it became a Roman
provincial town, this is an interesting and to a considerable degree a satisfying plan.
It shows the Canopic Way running from the Canopic (Canobic) Gate to the Necropolis
beyond Rhacotis (from northeast to southwest). The Canopic Way is cut at right angles
by the great crosstown street which extends from the Sun (Helios) Gate off Lake
Mareotis to the Moon Gate near the Heptastadion. To the extreme northeast beyond
the city walls is the Hippodrome, an enclosure for chariot racing with tiers of seats

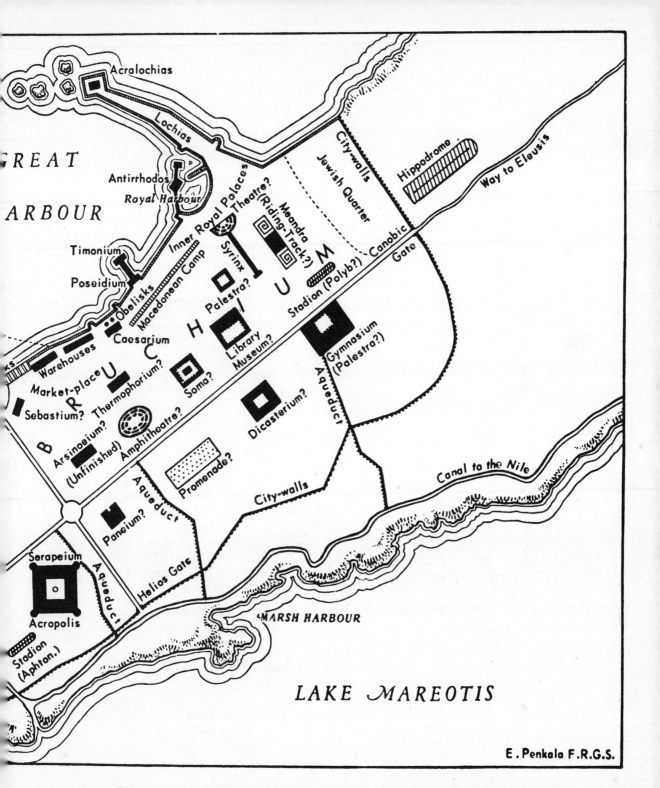

for the spectators, which was famous throughout the East. The Jewish quarter is southeast of Lochias. The large Brucheum quarter borders on the Jewish quarter and ends at the crosstown street running from the Sun Gate to the Moon Gate. Rhacotis fronts on Eunostos Harbor (to the northwest) and extends toward Lake Mareotis; in this quarter is the Serapeum (Serapeium). The Soma and the museum and library are placed close together, north of the Canopic Way but near it and at a considerable distance from the waterfront.

THE EARTH IS RECENTERED

The Emporium (market place), Apostases (warehouses) and Navalia (docks) lay west of the Timonium along the seafront as far as the mole.

Behind the Emporium rose the Caesareum, a stupendous temple which Cleopatra began in honor of Antony. After they took their own lives, Octavian finished it in honor of himself. He was worshiped there as Caesar Augustus, and the temple remained in imperial possession until Christian times.

A pair of obelisks, made in the time of the Egyptian pharaohs and brought from Heliopolis to Alexandria, stood beside this tem-

Mosaic depicting a seaport with two harbors. Approximately 6 feet long, the mosaic panel was one of a number found in pairs in the excavation of Cenchreae, an ancient town on the Saronic Gulf southeast of Corinth. The picture is made of pieces of glass of various colors and shapes assembled so as to form a double harbor lined with buildings; behind these appear the roofs of other buildings. Different kinds of ships, several fish, and a fisherman fill the space among them. Although the specific interpretation of these scenes is uncertain, the architectural panorama shown may give some

ple. By and by they became known as "Cleopatra's Needles." These obelisks—upright four-sided pillars, gradually tapering as they rose, and terminating in pyramids—were subsequently removed. One is now to be seen in New York's Central Park, the other on London's Embankment, the elevated bank with esplanade and gardens on the left side of the Thames.

The Gymnasium (sports grounds) and Palaestra (wrestling school) were both inland; they were near the Canopic Way in the eastern half of the town.

There was also a temple of Saturn, but its site is unknown.

idea of the city of Alexandria. Dating from the middle of the fourth century A.D., the panels, evidently packed in crates for shipping, were presumably to be mounted on walls as decorations. They are very difficult to photograph, and the accompanying drawing (*below*) shows the details of the mosaic more clearly. The excavation of Cenchreae was conducted by the University of Chicago and Indiana University for the American School of Classical Studies at Athens.

The mausoleum of Alexander the Great (the Soma) and those of the Ptolemies were in one enclosure near the point of intersection of the two main streets, and the museum with its library and theater for lectures and readings was in the same region. The library may have been damaged, or many books may have been lost, in 48 B.C. when Julius Caesar set fire to the Egyptian fleet in the harbor nearby. A few years later, in 40 B.C., Antony is said to have tried to repair the loss by presenting to Cleopatra the library of Pergamum. This great and beautiful city of Asia Minor possessed a library second only to that of Alexandria.

The Serapeum stood in the western part of the city. It was set off to advantage with extensive columned halls and many works of art. This magnificent temple, with its large statue of a sitting figure beautified with precious metals, was sacred to the god Serapis— healer of the sick, miracle worker, a god who was superior to fate and who had the character of a deity of the underworld. Serapis, whose worship was introduced by the first Ptolemy, was depicted with a kindly bearded face. His head was crowned with a fertility emblem. At the right knee of the seated god was the three-headed dog Cerberus, while the god's upraised left hand held a scepter or staff.

The great lighthouse, the most conspicuous edifice of Alexandria, rose on the eastern part of the Pharos island. A temple of Hephaestus, god of fire, also stood on Pharos at the head of the mole.

Such was the city Strabo knew. And it is likely that Alexandria had not changed very much by the second century A.D., when Claudius Ptolemy—the only great astronomer of Roman times and the last important one in Alexandria—lived. Cities then did not alter as they do now, though there were obviously some changes.

20 ⚬ *Ptolemy Scans*
the Circling Stars

"I know that I am mortal and ephemeral, but when I scan the multitudinous circling spirals of the stars, no longer do I touch earth with my feet, but sit with Zeus himself, and take my fill of the ambrosial food of the gods."

This saying is included in the *Greek Anthology* and bears Claudius Ptolemy's name. But even if it is not his, it gives a good picture of the astronomer. He is glimpsed as a man elevated far above other men by his sublime purpose and tranquillity.

Ptolemy's work is traditionally associated with Alexandria, and there is no reason to suppose that he ever lived anywhere else. He himself says, "We make our observations in the parallel [the degree of latitude] of Alexandria."

The name "Ptolemaeus" indicates that he was an inhabitant of Egypt of Greek or Macedonian descent. At the same time, his first name, "Claudius," shows that he had Roman citizenship; this was probably the result of a grant to an ancestor either by the emperor Claudius or by Nero.

Although the exact dates of Ptolemy's birth and death are unknown, the time when he lived can be determined by his astronomical observations. His first recorded observation, an eclipse

117

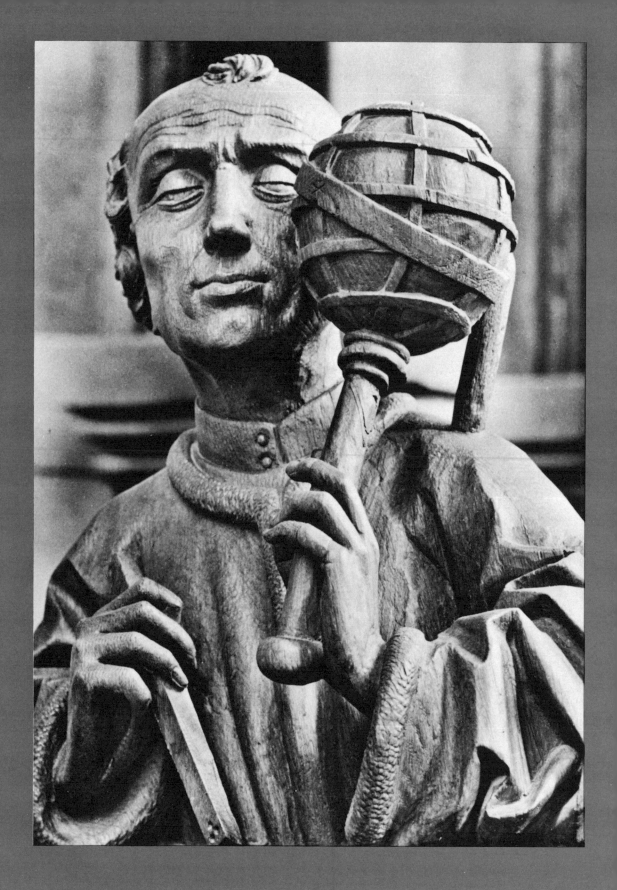

of the moon, was made in A.D. 127 and his last in A.D. 141. According to an Arabic story, the great astronomer lived to the age of 78, and an early annotator of classical texts stated that he was still alive in the reign of Marcus Aurelius which lasted from 161 to 180. From this evidence, it would seem that Ptolemy was probably born at the end of the first century A.D.

Ptolemy had the good fortune to live under the rule of some of the best Roman emperors. Trajan (emperor A.D. 98–117) was a soldierly ruler who conquered Armenia, Arabia, and Parthia and made these regions into Roman provinces. Strict and economical, he was also humane and progressive in his administrative policy. Hadrian (emperor 117–138) gave up some of Trajan's conquests, built a wall across Britain to hold back the wild highlanders from the north, and established a federal civil service. He also reorganized the army by introducing sound discipline, granted citizenship to many provincials, and encouraged learning. Antoninus Pius (emperor 138–161) improved the conditions of the slaves and embraced the principle that a man is innocent until proved guilty. Marcus Aurelius (emperor 161–180) was a Stoic philosopher and writer. His *Meditations* approached Christianity in spirit. However, he persecuted the Christians. He died while defending the frontier against the barbarians.

For two centuries, from the rule of Augustus (Octavian) to that of Marcus Aurelius, there was peace within the empire. The

THE EARTH IS RECENTERED

Roman Peace *(Pax Romana)* turned out to be a real blessing to the Mediterranean world, which had always been harassed by the horrors of war. Commerce and industry increased. Cities prospered. The refinements of life spread throughout the empire. The *Pax Romana*, however, described conditions only within the empire; on the frontiers, the Roman soldiers were constantly fighting.

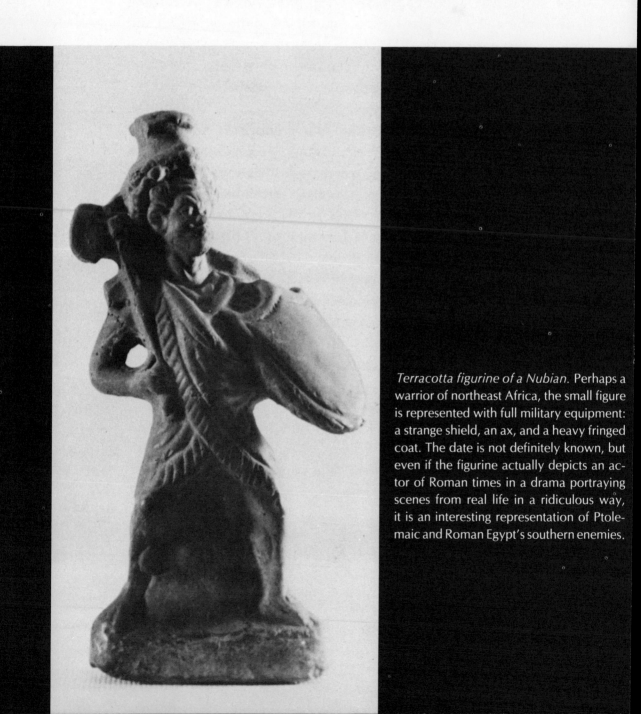

Terracotta figurine of a Nubian. Perhaps a warrior of northeast Africa, the small figure is represented with full military equipment: a strange shield, an ax, and a heavy fringed coat. The date is not definitely known, but even if the figurine actually depicts an actor of Roman times in a drama portraying scenes from real life in a ridiculous way, it is an interesting representation of Ptolemaic and Roman Egypt's southern enemies.

21 ✲ *The Thirteen Books of* the Almagest

Although Alexandria had declined considerably from its former greatness as a center of learning, the city still carried on a scholarly tradition in the time of Claudius Ptolemy. Living in Alexandria must have been a great advantage to him in his work and possibly in his education. At the very least, its libraries furnished him with important reference material. In the second half of the first century both libraries were very rich, although during the second century, when Ptolemy lived, the process of decline was rapid and it is believed that a large number of books were taken to Rome.

Both Euclid and Ptolemy composed textbooks which remained standard works in their particular fields for over a thousand years. Euclid, it will be remembered, is known primarily as a mathematician. His fame is based on the *Elements*. Ptolemy's achievement was far more complicated; he is especially renowned for the *Almagest* and the *Geography*. Both of these men were excellent teachers. Those who had preceded them had written descriptions of a particular thing or class of things or short presentations of a general subject. They wrote large works of encyclopedic nature; these were well organized and perfectly clear. The earlier books upon which their own were based were soon estimated to be incomplete and out

121

of date. The scribes stopped copying them and they disappeared.

Of Ptolemy's two great standard works, the best known is the *Almagest*; the title is that of the Arabic translation made about A.D. 800. Like Euclid's *Elements*, it consists of thirteen books. It is an account of all aspects of mathematical astronomy as the ancients imagined it. But it is more than that. A large part of the theory in the work is undoubtedly Ptolemy's own contribution. It is true that Ptolemy was not the kind of genius who produced new ideas; he took existing ideas and changed and extended them in order to obtain good agreement with observed facts. However, most scientists in all periods have worked in this way. As mentioned earlier, Ptolemy utilized Hipparchus's material with the utmost skill, but he also added to it an enormous body of facts based on observations of his own, not to speak of the unbelievable number of numerical calculations which underlie the tables of the *Almagest*.

The first two books of the *Almagest* are introductory; they explain astronomical assumptions and mathematical methods. Book 3 deals with the length of the year and the motion of the sun. Books 4 and 5 are mainly on the length of the month, the theory of the moon, and sizes and distances of the sun, moon, and earth. Book 6 deals with eclipses of the sun and moon. Books 7 and 8 deal with the fixed stars and the precession of the equinoxes. Books 9 through 13 are devoted to the motions of the planets.

The greater part of Books 7 and 8 is made up of the "Star Catalog." This is the earliest catalog of stars to have survived. It is a list of 1022 stars arranged under 48 constellations, giving each star's longitude, latitude, and magnitude (the brightness of a star expressed on a scale in which the brightest are numbered 1 and the faintest visible to the naked eye 6). To compile this catalog entirely from personal observation would have been a huge undertaking,

Claudius Ptolemy with a cross-staff. If he lacked telescopes, this ancient astronomer was by no means without instruments. André Thevet has shown him with the cross-staff, which was widely used in astronomical work and navigation at the time. In this instrument, the cross member, which was equipped with sights at each end, was gripped by the slide-joint at the exact center.

To find the angle subtended by two stars or the elevation of a body above the horizon or other reference point, Ptolemy applied his eye to the one sight on the long member of the instrument and brought the two objects into view through the sights at either end of the cross-staff. He read the angle from a graduated scale along the main member of the instrument.

and Ptolemy has frequently been denied credit for it. However, according to the modern view, there is no reason to disbelieve Ptolemy's claim that he made his own measurements, which he then compared with those of Hipparchus and earlier stargazers. Moreover, it is probable that this was the first time a star catalog was composed in this form.

The last five books of the *Almagest*, dealing with planetary motions, are the most famous of Ptolemy's works. The astronomers Eudoxus of Cnidus (408–355 B.C.) and Callippus of Cyzicus (about 370–300 B.C.) had imagined the planets to be attached to a complicated system of moving spheres. Ptolemy replaced these spheres with a system of moving circles. In his system, a planet revolves uniformly in a circle (epicycle), the center of which revolves uniformly on another circle (deferent), round a point not agreeing in position with the earth (eccentric). Ptolemy dealt with epicycles, deferents, and eccentrics in order to represent geometrically the principal irregularities of the motions of the planets, especially the stationary points and retrogradations—the backward motions of planets.

So Ptolemy refused to accept the ideas of Aristarchus of Samos, just as Hipparchus had done before him. They rejected these ideas because they did not agree sufficiently with the observations. Although Ptolemy's system was marked by error, at the time that it was stated formally for consideration it may have been more useful than the truth. People in those days were more concerned with the apparent than with the real planetary motions. The system provided a description of these which was almost accurate and could be understood by those for whom it was intended.

The earth-centered picture of the universe proved to be extremely attractive to medieval theologians. To them man and consequently his habitat, the earth, were the main purpose of creation. In fact, the Ptolemaic System, as the solar system centered upon the earth is called, gained complete and paramount influence over men's minds for more than 1400 years.

Head of Antoninus Pius. The Roman emperor is shown on the obverse of the coin depicting Serapis and the Pharos lighthouse (see page 66). He enjoyed a remarkably peaceful and prosperous reign; literature was encouraged and trade and communications were advanced. Claudius Ptolemy was a contemporary of this sovereign.

22 ⁑ *The Future Is Explored in the Heavens*

Ptolemy's *Geography*, in eight books, is a remarkable factual and scientific achievement. It is an attempt to map the world as it was known at the time. The main body of the work consists of the longitudes and latitudes of principal cities and of geographical features such as rivers. There must originally have been maps as well, but Ptolemy also included all the information needed for the drawing of accurate maps. Ptolemy's knowledge was drawn primarily from Marinus of Tyre, who was active not long before him, but the *Geography* included contributions of his own. He was probably the first to use systematic listings by latitude and longitude. The knowledge of the world presented in the *Geography* is often inaccurate, but its range and variety are nonetheless amazing.

Ptolemy's *Optics* has survived in a twelfth-century translation from Arabic into Latin. The *Optics* was divided into five books, but Book 1 and the end of Book 5 are lost. Book 5 deals with refraction —the change of direction of a ray of light when it passes from one medium to another of different density, as it is expressed in modern language. The phenomenon of refraction is demonstrated by the experiment of the coin in the container which appears to move to another position when water is poured in. This experiment is at least as early as Archimedes. Another experiment is included to determine the degree of refraction from air to glass, from water to glass, and from air to water. Ptolemy goes on to discuss astronomical refraction. He saw that rays of starlight would be bent as they passed from the thin air in the upper atmosphere to the denser air beneath. This causes a star to appear more directly overhead than it actually is. In this way, for example, the sun, moon, and stars remain in view after they have actually gone down below the horizon.

It is hard to fix the value of Ptolemy's contribution to the *Optics* because so little remains of the work of the scientists who

Drawing of the hemicycle of Alexandria. The hemicycle, an ancient form of sundial, was hollowed out of a block of stone, and the face was cut away from above at an inclination parallel with the plane of the equator. Thus the useless part of the hemisphere was removed, and the inclination of the dial would correspond with the latitude of the place for which it had been built. There were as a rule eleven hour lines, dividing the daytime into twelve hours. These, in most of the dials which have been found, are crossed by three parallel arcs, which mark the equinox and the summer and winter solstices. The pin projecting the shadow used as an indicator was horizontal. A dial of this kind, probably belonging to the Roman period, was found near Alexandria in 1852 at the base of Cleopatra's Needle. It is hollowed out of a block of stone 16½ inches high, 17 inches wide, and 11 inches deep. The corners have been broken off, but most of the hour lines can be seen, with the Greek letters ΑΒΓΔΕΣΖΗΘΙ numbering them. The face of the block is seemingly inclined to correspond with the latitude of Alexandria, and the base is cut in six small sloping steps.

Life in the city. Four phases of everyday life are depicted by these objects. The terracotta lamp *(top left)* is in the form of an elderly man. Wearing a hooded cloak, he has a pointed beard and is holding a lantern in his right hand and a ladder in his left; the lower part of the object is missing. The pottery apparently represents a lamplighter. While there is no evidence in literature or in papyri for lamplighters in Alexandria, such figures are well known in the Serapeums of Egypt and Athens. The dwarf playing the organ *(top right)* is another lamp; it was found

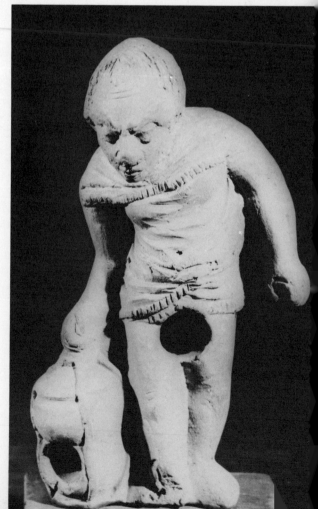

in the Faiyum, a province in northern upper Egypt. Still another lamp *(bottom left),* perhaps of Roman times, is in the form of a slave carrying a lantern. The slave with a lantern, standing or seated, is a common type; generally the slave is black. The old woman *(bottom right)* is holding a flagon in her left hand; a similar statue with the Greek inscription ''nurse'' on the base has been recovered. This object also probably dates from Roman times, but it has Hellenistic sources.

129

preceded him in this field. The most striking feature of the *Optics* is the method of establishing theory by experiment, which often included building special apparatus. Whether the text is drawn primarily from other sources or is a product of Ptolemy's own thought, the work is an impressive example of the development of a mathematical science with appropriate regard to the physical data and is worthy of the author of the *Almagest*.

Among Ptolemy's other works is the *Tetrabiblos*, a textbook in astrology in four books which he considered to be a natural addition to the *Almagest*. The *Almagest* makes it possible to predict the positions of the celestial bodies; the *Tetrabiblos* sets forth the theory of their influence on terrestrial things. Astrology should not be condemned because it sometimes fails, Ptolemy argued, no more than a physician should be condemned who fails to cure a patient, or the art of navigation because of shipwreck.

The *Tetrabiblos* is made up of material gathered from Chaldean, Egyptian, and Greek folklore and from earlier writings, though the author frequently develops his own ideas. It endured as a standard work until modern times because it is so complete and so well organized. It explains the technical concepts of astrology and deals with predictions of a general nature (such as wars, earthquakes, and the weather) and with predictions relative to individuals (such as marriage, foreign travel, and material fortune). Horoscopes —diagrams of the heavens, showing the relative positions of planets and the signs of the zodiac, for use in calculating influences at birth, foretelling events in a person's life, and so on—are also discussed.

Astrology was accepted by nearly everyone in the Roman empire, and even men of genius like Ptolemy could not escape its sway. For Ptolemy was a man of his age and of the land in which he lived.

PART FIVE ✱✱✱ THE FALL
OF THE CITY

23 ✩ *The Murder of*
the Alexandrian Youth

In A.D. 215 the Roman emperor Caracalla visited Alexandria. His real name was Marcus Aurelius Antoninus; Carcalla was a nickname derived from the long hooded tunic or coat worn by the Gauls, which he chose as his favorite dress after he became emperor.

Shortly after becoming joint emperor with his brother Geta, Caracalla murdered him, along with a large number of the most eminent men of the state. He was lavish in expenditure as well as cruel, and it was rumored that he wanted to marry his mother.

Now the Alexandrians were well known as a people of free speech. They had a keen sense of humor, and this frequently took the form of bitter satire with which they entertained themselves in the theaters as well as in the streets. The Alexandrians had made references to Caracalla's monstrous personality in their satires, and the emperor repaid them for their insults when he visited them.

This is how he did it: he commanded his troops to put to death all youths of the city able to bear arms. The order appears to have been carried out even beyond its precise wording; there was a general massacre, and the city was plundered. In spite of this crushing misfortune, the resilience of Alexandria was such that the city shortly regained its old splendor and for some time was rated the first city of the world after Rome.

133

Figurine of a fettered black slave. The fine small bronze Hellenistic figure of a black slave or prisoner with hands chained behind his back was found in a fertile region near Cairo. He is looking up with a defiant smile. Slaves formed a large part of the population of Alexandria.

THE MURDER OF THE ALEXANDRIAN YOUTH

During this time, Alexandria gained fresh importance as a center of church government and Christian theology—the branch of religious science that deals with God. It was in this city that the doctrine of Arianism was formulated, and it was in this city also that Athanasius, a lifelong opponent of Arianism, worked and triumphed.

A basic creed of Christianity is that the Divinity is made up of three Persons in one substance: the Father, the Son, and the Holy Spirit. The ecclesiastic Arius tried to apply logic to mystery. He concluded that the Son was not of the same divine substance as the Father. Athanasius acted as the champion of orthodox faith as set forth at the Council of Nicaea called by Constantine the Great in 325. He became the patriarch of Alexandria in 328 and was persecuted whenever the believers in Arianism got the upper hand. This father of the church was exiled five times, but each time he was restored, and during his last years he continued his labors at Alexandria.

24 ✦ A Library Is Destroyed

As native Egyptian influences began to re-establish themselves in the Nile Valley, Alexandria gradually became an alien city, more and more separated from Egypt. When the peace of the empire ended during the third century A.D., it lost a great deal of commerce and diminished rapidly in magnificence and population.

In A.D. 273 the city suffered another major loss under the rule of Aurelian. Aurelian was the son of a peasant-farmer on the estate of a Roman senator. By merit he rose from the rank of a common soldier to become emperor.

In 272 Firmus, a wealthy merchant who seems to have had connections with Ethiopian and Saracen tribesmen, marched on Alexandria, led astray the changeable masses, and seized the government of Egypt. He was a huge man of great strength who could eat a whole ostrich in a day and drink more than Aurelian's generals and yet remain sober.

Aurelian moved at once against this enemy and crushed the revolt. He surrounded Firmus with an armed force in the Brucheum quarter, and the destruction wrought there was great. Firmus, whom he regarded as a robber, was captured and crucified.

As mentioned before, the museum and its library were situ-

Pompey's Pillar. So called because of the medieval legend that it marked the tomb of Pompey the Great, the imposing column still stands upon the shapeless hill where the Serapeum was situated; the Arabs call it simply al-Amūd (post, column). The substructure is constructed of blocks taken from older buildings. On the eastern face there is a block of green granite with a Greek inscription in honor of Arsinoë II. On the opposite face, upside down in a recess, are the figure and hieroglyph of Seti I, who ruled Egypt from 1313 to 1292 B.C. This suggests the great age of the settlement on Rhacotis. On the western side of the granite base of the column, about 10 feet up and invisible from the ground, there is a four-line Greek inscription: "To the most just Emperor, the tutelary God of Alexandria, Diocletian the invincible: Postumus, prefect of Egypt."

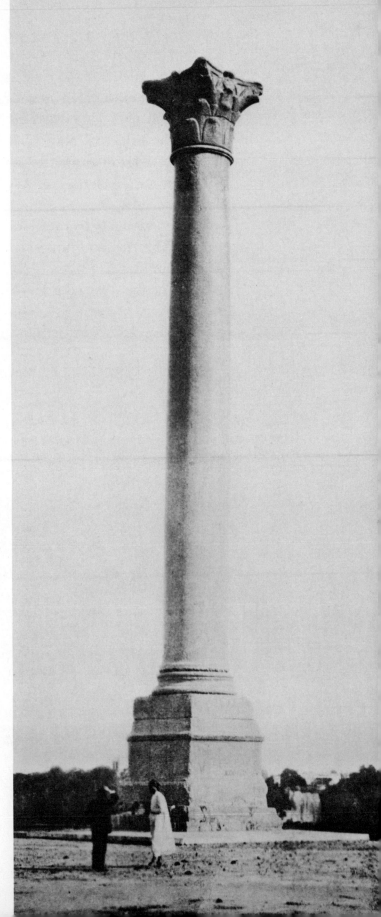

THE FALL OF THE CITY

ated in the Brucheum quarter, which had for so long been the home of distinguished men. It is very possible that the library perished when this district was destroyed. The library in the Serapeum then became the principal library of Alexandria.

A few decades later, in 296, a fresh revolt broke out in Egypt. The Emperor Diocletian, whose parents had been slaves, subjected Alexandria to a siege of eight months. Thousands of the citizens were killed savagely by the soldiers, and a large part of the city was burned. The Alexandrians set up Diocletian's statue on top of an old column that had been used by a Roman prefect as a signal for sailors. The famous red granite landmark, 90 feet high and known for centuries to travelers as Pompey's Pillar, was erected at the time to recall the punitive discipline of Diocletian.

Small statue of Attis on a lion's back. Attis was a youth of Phrygia who was loved by Cybele. In a jealous rage, this great nature goddess caused him to go mad. He castrated himself and died, and the goddess then begged Zeus to preserve his body free of decay. The bronze object is part of the find of Galjûb (see page 90).

25 ⁜ The Serapeum Is Plundered

In the time of Theodosius I the Christian Church triumphed, and paganism was officially forbidden throughout the empire.

In 389 or 391 an edict of this Roman emperor ordered the destruction of the Serapeum. Theophilus, then patriarch of Alexandria, is said to have led the crowds into the temple to destroy it himself. He has been described by the eighteenth-century English historian Edward Gibbon, in *The History of the Decline and Fall of the Roman Empire*, as "the perpetual enemy of peace and virtue; a bold, bad man, whose hands were alternately polluted with gold and with blood." The library was pillaged by the Christians, although it is possible that the destruction was incomplete and that some of the books were rescued in one way or another.

During this period there lived in Alexandria a teacher in the museum named Theon, who was an astronomer and mathematician; he wrote an elaborate commentary, or series of comments, on the *Almagest* and edited Euclid's *Elements*. Another illustrious Alexandrian was Theon's gifted daughter Hypatia, who taught at the museum as well. She was the first woman mathematician and wrote commentaries, which are now lost, on the *Conics* of Apollonius and the *Arithmetica* of Diophantus. She also assisted her father with his writings.

139

The Serapeum. The most beautiful of that city's temples is depicted on the reverse of this bronze coin minted in Alexandria in the time of Trajan. The temple contained many marvels, including a colossal statue of Serapis and supposedly a window that opened to the east and caused the first ray of the sun to fall on the mouth of the god.

Hypatia was a philosopher as well as a mathematician. Not much is known of her philosophical opinions, but she seems to have adopted the intellectual rather than the mystical side of Neoplatonism and to have been a follower of Plotinus.

About A.D. 400 Hypatia became the head of the Neoplatonic school in Alexandria. Her learning was so great that she was superior to all philosophers of the time. She was eloquent, modest, and beautiful. People wishing to work in philosophy came from all parts of the world to study with her. The most famous of her pupils was Synesius of Cyrene, who was converted to Christianity in middle age and became a bishop.

From the time of the destruction of the Serapeum, there was a constant commotion between the Christian zealots, the Jews, and the pagans. Cyril, Theophilus's nephew, became patriarch of Alexandria in 412. Trouble arose between the Jews and Christians, and Cyril wanted to drive out the Jews altogether. The prefect Orestes opposed this. Frequent conflicts occurred. Mobs paraded the streets, which were no longer safe. Hypatia was accused of undue familiarity with Orestes, and the clergy believed she obstructed the friendship of Orestes with their archbishop, Cyril.

There is a saying, "when the mob goes down the street, God goes inside." So it was in March of 415. As this beautiful philosopher was driving in her chariot, a fanatical Christian mob incited by Cyril dragged her from the vehicle and into the Caesareum, which had become a church. There she was stripped naked, her flesh was torn off her bones with sharp oyster shells, and she was finally burned. Hypatia was one of the earliest martyrs of science.

As the fourth century was nearing its end, no new scientific work was being done in Alexandria. Any original thoughts produced by the Alexandrian school were philosophical speculations. The principal work of its scholars was commenting, editing, and recounting the glories of an age gone by.

26 ✲✲ Alexandrians in Athens and Constantinople

The Brucheum and the Jewish quarters were abandoned in the fifth century, and the central monuments, Alexander's mausoleum and the museum, were in ruins. On the mainland, life appears to have concentrated in the neighborhood of the Serapeum and the Caesareum, both of which had become Christian churches. But the Pharos and Heptastadium quarters were unimpaired and continued to contain many inhabitants.

At this time, some Alexandrians migrated to Athens. There Plato's Academy still carried on a weakened existence—a small center of paganism while all around it Christianity became more widely accepted. The Academy had ceased to be an advanced school of mathematics. Most of the teachers and students were concerned with Neoplatonic arithmetic—that is, number mysticism. Proclus was one of the last heads of the Academy. A Neoplatonic philosopher, he is looked upon as the last of the great teachers of this school of thought. Born in Constantinople in about 410, he studied at Alexandria but too late to be a student of Hypatia. He vigorously defended paganism and opposed Christianity and was head of the Academy until his death in 485. At last, in 529, the Christians won over Justinian the Great and got that emperor to forbid the study of all "heathen learning" in Athens. The school of Athens died in its turn.

Other Alexandrians migrated to Constantinople. This city on the Bosporus had originally been called Byzantium. Constantine the Great chose it as the site for a new Christian capital of the East and rebuilt it, creating what was virtually a new city, with the same enthusiasm and attention to details Alexander had shown in Egypt six centuries earlier. In 330 the city was officially refounded by the emperor and renamed Constantinople (City of Constantine). It is now called Istanbul. Learning did not develop to great heights in Constantinople, but the city became a center of culture in the East. It remained so for about 1,000 years, until the Turks captured it in 1453. Although the scholars of Constantinople did not add much to the world's accumulation of knowledge throughout these centuries, they guarded from destruction a great deal of learning.

Oribasius, who was born in Pergamum about A.D. 325, was a physician to the Byzantine court in Constantinople and the personal physician of the Emperor Julian. Julian publicly announced his conversion to paganism in 361 and consequently is known as Julian the Apostate. The emperor encouraged his physician to write a huge medical encyclopedia. It consisted of seventy books, but only one-third of it has survived. Oribasius's *Medical Collection* helped to preserve many earlier writings which would have been lost in other circumstances. He quoted frequently, and he always referred these quotations to their authors. The medical experience and knowledge, essentially of pagan origin, that was available in the second half of the fourth century is contained in this great work. The Christian emperors who followed Julian the Apostate persecuted the pagan physician, but he was finally recalled to Constantinople and permitted to continue his work. He died in about 400.

Mosaic of Alexandria and the Pharos. The sixth-century floor mosaic was found in the north side of a church at Jerash (ancient Gerasa) dedicated to St. John the Baptist in A.D. 531. Gerasa was a city of the Decapolis, a region in the north of Palestine, which flourished in the second and third centuries A.D. On the left are some of the palms with bunches of red dates for which the north of Egypt is still well known. The large group of buildings represents the walled city of Alexandria with its name in Greek above, and on the right is the tower and part of the octagonal portion of the famous lighthouse.

144

One theological conflict of the Byzantines actually turned out to be beneficial to learning. It was believed then, as it is today, that there are in Christ two natures (human and divine) but one person. Nestorius, who was appointed patriarch of Constantinople in 428, preached the doctrine that there are in Christ two natures and two persons. He was deposed for heresy by the Council of Ephesus in 431 and banished to the Libyan desert. The followers of Nestorius were persecuted and moved eastward. First they went to Edessa (the modern Urfa) in Mesopotamia, later to Baghdad and elsewhere. There was a medical school in Edessa, and in this scientific community they translated into Syriac—which had become the common language of western Asia—the works of Aristotle, Plato, Euclid, Archimedes, Heron, and Ptolemy and many other books of philosophy and science. These Syrian books were later translated into Arabic.

So these refugees helped to develop Arabic science. In the course of time, the works were translated from Arabic into Latin, Hebrew, and modern Western languages. Ancient science reached us through that roundabout way, and gratitude must be expressed both to the men who found out about the universe and to those who carried this knowledge from one land to another and thus helped to shape the modern mind. For science today is but the continuation and fulfillment of the science of yesterday and could not have come into being without it.

27 ✢ *Books Are Burned in the Baths*

Khosrau II, called Khosrau Parvez (the victorious), was king of Persia from 590 to 628. He had been helped in securing his throne by Mauricius, emperor of the Eastern Roman Empire. In 602 the Roman centurion Phocas was raised to emperor by mutinous soldiers and had Emperor Mauricius killed. Khosrau then made war on the Eastern Roman Empire to take vengeance for Mauricius's death. He conquered most of the regions of southwestern Asia, occupied Egypt without conflict, and reached Chalcedon opposite Constantinople. He took Alexandria in 616.

Khosrau was finally defeated and driven back by Heraclius. Heraclius was the successor of the usurper Phocas, who had been overthrown, tortured, and beheaded. After ten years of Persian rule, the success of this great emperor restored Egypt to the empire, and for a time it once more had a governor.

Then, in 639, Omar I, the second of the caliphs that ruled Arabia after the death of Mohammed, sent his general Amr ibn-al-As, whose name is sometimes given as Amru in English, to invade Egypt. Victory followed victory for the Arab invaders. Alexandria was surrendered to Amru in November 641 on the condition that it should be occupied by the Moslems in September of the following

147

year. The ease with which this valuable province was pulled away from the Roman empire seems to have been due to the incompetence of the Roman generals and the violation of allegiance by the governor of Egypt, Cyrus, who was also patriarch of Alexandria. Cyrus's motives for surrendering Alexandria, and with it all of Egypt, have been interpreted in different ways. Some have believed that he was a secret convert to Islam, the Mohammedan religion.

Amru was able to write to the caliph Omar that he had taken a city containing "4,000 palaces, 4,000 baths, 12,000 dealers in fresh oil, 12,000 gardeners, 40,000 Jews who pay tribute, 400 theaters or places of amusement."

The final end of the Alexandrian school came in 642 when the Mohammedans occupied the city and destroyed the remainder of the great library. The famous story of the final destruction of the library is told by Barhebraeus (the form of his name commonly used in English is Abulfaraj) in his work, *History of the Dynasties.* Abulfaraj, whose father was a Jewish physician who became a Christian, was a churchman and the last classical author in Syriac literature, and in the thirteenth century he wrote this description of the destruction of the books:

"John the Grammarian, a famous peripatetic philosopher, being in Alexandria at the time of its capture, and in high favor with Amr [Amru], begged that he would give him the royal library. Amr told him that it was not in his power to grant such a request, but promised to write to the caliph for his consent. Omar, on hearing the request of his general, is said to have replied that if those books contained the same doctrine with the Koran [the Mohammedan sacred scripture], they could be of no use, since the Koran contained all necessary truths; but if they contained anything contrary to that book, they ought to be destroyed; and therefore, whatever

148

their contents were, he ordered them to be burnt. Pursuant to this order, they were distributed among the public baths, of which there was a large number in the city, where, for six months, they served to supply the fires."

The historical accuracy of this account of the destruction of the library by the Arabs is doubtful. When the disordered conditions of the times and the neglect into which literature and science had fallen are considered, it is not difficult to believe that there were very few books left to be destroyed by the soldiers of Amru. But the question is not easy to settle, and a large number of names could easily be mentioned on either side of the argument. At the same time, some of the most careful investigators confess the difficulty of making a decision.

In 645 an attempt was made with a force under Manuel, commander of the imperial forces, to regain Alexandria. Amru quickly returned with a large army and stormed the city in 646. He killed the fighters, destroyed the walls of the city, and carried away the children as captives.

This is the end of the story. Alexandria now rapidly declined in importance. In the early part of the seventeenth century, in *Relations of Africa*, George Sandys, a Christian traveler, said of Alexandria: "Queene of Cities and Metropolis of Africa: who now hath nothing left her but ruines; and those ill witnesses of her perished beauties: declaring that Townes as well as men, have their ages and destinies."

Postscript

In the early eighteenth century Alexandria had become a desert and a robbers' nest. There was little to remind one of the great imperial metropolis of nearly a million people under the rule of the first Roman emperor Augustus.

After a long period of decline in significance because of the rise of Cairo and the failure to dredge the silted harbor, the city revived commercially when it was joined to the Nile by a canal in the early nineteenth century. The modern city is built on the strip of land which separates the Mediterranean from Lake Mareotis and on a peninsula connected with the mainland by an isthmus which was originally the mole leading to Pharos island. Over the centuries this mole has become blocked up with sediment. It is now an isthmus half a mile in width. The point of land extending into the sea at the western end of the peninsula is Ras-et-Tin, the Cape of Figs. The eastern cape is called Pharos or Kait Bey. There are two harbors, east and west.

The population of Alexandria today is about twice

The Alexandrian obelisk. Taken in 1879, the photograph recalls the feeling of disgust aroused by the immediate surroundings of this monument before it was removed to New York. No one considered the obelisk worthy of care and protection. Two businessmen broke pieces from the angle of the shaft and the edges of the intaglios and sold them to relic hunters. Strangers, who rarely spent more than a few seconds examining it, departed hastily because of the disagreeable odors and clamors for *backsheesh* (Arabic, gift). Nothing was more neglected and less appreciated than was the Alexandrian obelisk by the residents of Alexandria and the tourists who passed through the city en route to the Nile.

what it was in the days of Augustus. It still maintains the cosmopolitan spirit brought into existence by its founder, Alexander the Great, over 2,000 years ago. Among its inhabitants are Greeks, Egyptians, Jews, and Europeans. Its industries range from the ancient art of rug weaving to paper manufacture.

There is a lesson to be learned from ancient Alexandria. Today the world is once again in a condition of disorder and confusion, and it is a real possibility that modern technological civilization will not endure and that its great cities will be buried in sand or engulfed by forests. In the past man acted in response to such a threat by flight, carrying with him vestiges of his culture, which, as has been pointed out in the case of Alexandria, survived in part to be incorporated in later civilizations. Today, however, there are not many refuges left on the planet earth, and escape into space, if feasible at all, could be accomplished only by a few. If modern civilization is to survive, different methods are required for its preservation.

Aerial view of modern Alexandria, showing the double harbor.

GLOSSARY
SELECTED BIBLIOGRAPHY
ACKNOWLEDGMENTS
NAME INDEX

Glossary

In the field of lexicography, Zenodotus of Ephesus, the first critical editor of the *Iliad* and the *Odyssey*, is known for his *Homeric Glossary*. He often relied on guesswork to give the meaning of difficult words, but he opened the way for the scholarly study of language. He also compiled a dictionary of foreign words.

The following glossary includes foreign words, words and phrases used in this book which are not in the common vocabulary, and fuller explanations of some subjects discussed in the text.

Alexandrian library. The library begun at Alexandria by Ptolemy I and largely collected by Ptolemy II. There were two libraries at Alexandria and they were the most famous of the ancient world: the larger, or "mother library," in the Brucheum quarter, was connected with the museum; the smaller, or "daughter library," was placed in the Serapeum.

Alexandrian museum. A kind of academy founded at Alexandria by Ptolemy I and developed mainly by Ptolemy II. It was primarily an institute for scientific research and illustrated, with the library, the greatness of Alexandrian culture.

Alexandrian school. The pagan school of literature, science, and philosophy which flourished at Alexandria while that city was ruled by the Greeks and the Romans.

Arianism. The doctrines of Arius, denying that Jesus was of the same substance as God and holding instead that he was only the highest of created beings.

armillary sphere. An ancient astronomical machine, or skeleton sphere, formed by putting together a collection of rings, designed to represent the positions of important circles of the celestial sphere. It turns on its polar axis within a meridian and horizon.

157

GLOSSARY

astrology. The pseudoscience which deals with the influence of the stars upon human affairs and with foretelling terrestrial events by their positions and aspects. It commanded the allegiance of some of the best minds of the ancient world, including Hipparchus and Claudius Ptolemy.

Athanasianism. The teaching of Athanasius, particularly that the Son is of the same substance with the Father, in opposition to Arianism.

aureus. A gold coin of ancient Rome.

bibliography. The history or description of books and manuscripts, with notices of the editions, the dates of printing, and so forth.

caliph. A title of the successors of Mohammed both as temporal and spiritual rulers.

chiton. A gown or tunic with or without sleeves. It was worn by both men and women in classical times, usually next to the skin.

chronology. The science which deals with measuring or computing time by regular divisions or periods and which assigns to events their proper dates.

cross-staff. An instrument for measuring the angle of elevation of celestial objects. It consists of a calibrated staff with another shorter staff perpendicular to and sliding on it.

Delphic oracle. The famous shrine of Apollo which was venerated by all Greeks of the ancient world. Here it was believed that the god, through his priestess, the Pythia, foretold the future. Many of the replies of the oracle were ambiguous, giving rise to the phrase "delphic response"—that is, an answer that can be interpreted in two ways.

drachma. The basic unit of ancient Greek coinage in either silver or gold.

dynasty. A race or succession of kings of the same line or family.

158

geocentricism. The belief that the earth is the center of the universe.

Gnosticism. The doctrines of certain early Christian sects, which embodied substantial elements of Oriental mysticism and Greek philosophy. Gnosticism commonly involved a plan of salvation through the acquisition of secret knowledge (gnosis).

grammarian. One versed in the science of languages; a philologist.

heliocentricism. The theory which places the sun at the center of the planetary system.

Hellinic culture. See *Hellenism.*

Hellenism. Greek character, spirit, or culture, especially the type of culture represented by the ideals of the classical Greeks, as in their regard for athletic vigor and grace, cultivation of the arts and sciences, devotion to civic social organization, and social and ethical attitudes. In the technical language of modern classical scholarship, Hellenism refers exclusively to Greek culture in the *Hellenistic age.*

Hellenistic age. The period of Greek history which began with the rule of Alexander the Great in Asia and which saw an integration of Oriental and Hellenic cultures, producing an international culture different from either.

hemicycle. A sundial in the form of a concave quarter sphere. It has a rodlike gnomon lying within one radius and is marked on its surface with arcs which lie in the same plane as the gnomon.

horoscope. A diagram representing the twelve houses of heaven and showing the relative positions of planets and signs of the zodiac by which astrologers profess to tell beforehand the events of a person's life.

hyparchus. A lieutenant governor.

Judaism. The religious doctrines and rites of the Jews, as reflected in the Old Testament and the Talmud.

159

GLOSSARY

magnitude. A degree of brightness of a heavenly body on a selected scale. In his star catalog, Claudius Ptolemy divided the stars arbitrarily into six magnitudes, in which the brightest were numbered 1 and the faintest visible to the naked eye 6.

mole. A mound or massive work formed of masonry or large stones laid in the sea, often as a structure to protect a harbor or beach from the force of waves.

Neoplatonism. A philosophical and religious system developed at Alexandria in the third century A.D., based on the doctrines of Plato and other Greek philosophers, combined with elements of Oriental mysticism and some Judaic and Christian concepts.

obelisk. An upright four-sided pillar, gradually tapering as it rises and ending in a pyramid. It is ordinarily a single great stone. Egyptian obelisks are generally covered with hieroglyphic writing (picture script) setting down the triumphs and other achievements of kings. Cleopatra's Needles, removed in ancient times from Heliopolis to Alexandria, are examples.

obliquity of the ecliptic. The angle between the planes of the earth's equator and orbit (ecliptic), the mean value being 23° 26′ 54″.21 in 1930 and decreasing 0″.47 per year. The rate of decrease will gradually slow down, and after many thousands of years, the obliquity will be about 22° 30′, after which it will start to increase.

obverse The side of a coin, medal, or the like that bears the principal design (opposed to *reverse*).

paganism. Pagan beliefs or practices. In modern usage, that is pagan which is not Christian, Jewish, Mohammedan, etc.; the word refers especially to past customs, sentiments, beliefs, or their survivals, and frequently implies contrast with Christianity rather than opposition to it.

papyrus rolls. Written scrolls made of papyrus. The use of papyrus for manuscript extended from the Pyramid Age to the fourth century A.D., with occasional use until the ninth century. It was prepared by cutting the pith of the papyrus plant

—which was sticky—into strips, laying the strips side by side so that they slightly overlapped, placing another layer of strips on top of and at right angles to the first layer, pressing the whole mass together into a homogeneous surface and allowing it to dry.

patriarch. In the early Christian church, any of the bishops of the ancient sees of Constantinople, Alexandria, Antioch, Jerusalem, or Rome with authority over other bishops.

Pax Romana. Roman peace. For two centuries, from the rule of Augustus (27 B.C.–A.D. 14) to the reign of Marcus Aurelius (A.D. 161–180), there was peace within the Empire.

pharos. A lighthouse or beacon to guide seamen. In ancient times, Pharos was an island in the harbor of Alexandria on which a famous lighthouse was located, and therefore the lighthouse received the same name. The Pharos was a symbol of Alexandrian prosperity and one of the seven wonders of the ancient world.

philology. The study of literature and of relevant disciplines.

polymathy. Wide and varied learning.

precession of the equinoxes. A slow westward motion of the equinoxes first discovered by Hipparchus and caused by the action of the moon and sun on the earth's equatorial bulge.

prefect. In ancient Rome, any of various high officials or magistrates placed at the head of a particular command, charge, or department. The function and rank of the prefects varied widely.

province. A country or region more or less remote from the city of Rome brought under Roman government.

Ptolemaic system. The system maintained by Claudius Ptolemy, which supposed the earth to be the fixed center of the universe about which the sun and stars revolved.

GLOSSARY

refraction. The deflection from a straight path undergone by a ray of light, heat, sound, or the like, in passing obliquely from one medium to another in which its velocity is different, as from air into water or from a denser to a rarer layer of air.

reverse. The side of a coin, medal, or the like that does not bear the principal design (opposed to *obverse*).

Roman Empire. The empire of the ancient Romans, established in 27 B.C. when Octavian was given the title Augustus Caesar. Under Diocletian, who became emperor in A.D. 284, the empire was divided into East and West, but it was only after the death of Theodosius I (A.D. 395), that the Western Roman Empire and the Eastern Roman Empire (called also the Byzantine or Greek Empire) were finally separated.

satrap. In ancient history, the name given by the Persians to the governors of the provinces. Egypt was a satrapy (the territory or jurisdiction of a satrap) under the Persian Empire, and when Alexander the Great conquered Egypt, he initially maintained the Persian satrapy administration. Thus Ptolemy became the satrap of Egypt.

skaphē. An improved sundial, invented by Aristarchus of Samos, consisting of a hemispherical bowl with a needle erected vertically in the middle to cast shadows.

stadium. A Greek measure of length, being the chief one used for itinerary (the route of a journey) distances. It was also adopted by the Romans for nautical and astronomical measurements.

stele. A burial stone.

tepidarium. In the ancient Roman baths, a warm room used for sitting. It was intermediate in temperature between the frigidarium and the caldarium.

theology. The study of the nature of God and religious truth.

Selected Bibliography

The first experiments in bibliography were made in producing catalogs of the Alexandrian libraries. This bibliography contains the most important books and articles used in writing *City of the Stargazers*. Great care was taken to rely on the best reference works and to exclude narratives which have been discredited. However, in dealing with a period as remote as that which has been described in this book, even modern scholarly sources disagree. What really happened is often uncertain, and the truth will probably never be known. For example, there is a great deal of doubt about the order of succession of the first librarians of Alexandria. Some believe that Eratosthenes was appointed head librarian on the death of Zenodotus. Others believe that Apollonius of Rhodes succeeded Zenodotus, and still others that Callimachus came after Zenodotus. This is a matter of interpretation of the available evidence. Moreover, future archaeological finds could modify the details of the story of Ptolemy and his museum and library. Thus the Polish excavations carried out in Alexandria in the past few years under Kazimierz Michalowski, director of the Polish Center of Mediterranean Archaeology in Cairo, will undoubtedly increase our knowledge of the ancient city, but these finds have not yet been published in the English language.

BERNARD, ANDRÉ. *Alexandrie La Grande*. Paris: Arthaud, 1966.
Dictionary of Scientific Biography. CHARLES COULSTON GILLISPIE, editor in chief. New York: Charles Scribner's Sons. "Aristarchus of Samos" by William H. Stahl, vol. I (1970); "Eratosthenes" by D.R. Dicks, vol. IV (1971); "Claudius Ptolemy" by G. J. Toomer (vol. to be published).

BIBLIOGRAPHY

DREYER, J. L. E. *A History of Astronomy from Thales to Kepler.* Formerly titled *History of the Planetary Systems from Thales to Kepler.* Revised with a Foreword by W. H. Stahl. New York: Dover Publications, Inc., 1953.

Encyclopedia Britannica. 14th ed. "Alexandria," "Alexandrian School," "Libraries."

FORSTER, E. M. *Alexandria: A History and a Guide.* Anchor Books. New York: Doubleday & Company, Inc., 1961.

GINGERICH, OWEN. "Musings on Antique Astronomy," *American Scientist,* 55:1 (1967).

HEATH, SIR THOMAS. *Aristarchus of Samos, the Ancient Copernicus: A History of Greek Astronomy to Aristarchus, Together with Aristarchus's Treatise on the Sizes and the Distances of the Sun and of the Moon* (a new Greek text with translation and notes). Oxford: The Clarendon Press, 1913.

HUTCHINS, ROBERT MAYNARD, editor in chief. Great Books of the Western World. Vol. 16: *Ptolemy, Copernicus, Kepler.* Chicago, London, Toronto: William Benton, 1952.

JEANS, SIR JAMES. *The Growth of Physical Science.* New York: The Macmillan Company, 1948.

NEUGEBAUER, O. "Notes on Hipparchus," in *The Aegean and the Near East, Studies Presented to Hetty Goldman,* edited by SAUL S. WEINBERG. New York: J. J. Augustin, Inc., 1956.

NEWELL, EDWARD T. *Royal Greek Portrait Coins.* Racine, Wisconsin: Whitman Publishing Company, 1937.

The Oxford Classical Dictionary. London: Oxford University Press, 1949.

PARSONS, EDWARD ALEXANDER. *The Alexandrian Library: Glory of the Hellenic World. Its Rise, Antiquities, and Destructions.* New York: The Elsevier Press, 1952.

ROSTOVTZEFF, M. *The Social & Economic History of the Hellenistic World.* 3 vols. Oxford: The Clarendon Press, 1941.

SARTON, GEORGE. *Ancient Science and Modern Civilization.* Harper Torchbooks/Science Library. New York: Harper & Brothers, 1959.

———. *A History of Science. Hellenistic Science and Culture in the Last Three Centuries* B.C. Cambridge: Harvard University Press, 1959.

———. *Introduction to the History of Science.* Vol. 1: *From Homer to Omar Khayyam.* Baltimore: published for the Carnegie Institution of Washington by the Williams & Wilkins Company, 1927.

SCHWARTZ, GEORGE, and BISHOP, PHILIP W. *Moments of Discovery.* Vol. 1: *The Origins of Science.* New York: Basic Books, Inc., 1958.

Acknowledgments

The author wishes to thank the following persons and institutions for assisting him in the picture research on this book: Adele Matthysse of the Burndy Library; Beulah F. Shonnard and Nancy Waggoner of the American Numismatic Society; The New York Public Library; and the Wilbour Library of Egyptology, The Brooklyn Museum.

Grateful acknowledgment is made to the individuals, institutions, and publishers that have given permission for the use of illustrations, as follows:

Alinari–Art Reference Bureau—page 99 *top*

Allard Pierson Stichting, Archaeologisch-Historisch Instituut der Universiteit van Amsterdam—page 120

American Elsevier Publishing Co., Inc.—pages 82, 112–113 (from *The Alexandrian Library: Glory of the Hellenic World* by Edward Alexander Parsons, New York, 1952)

The American Numismatic Society, New York—pages 36, 37, 66, 93, 125, 140

Anderson–Art Reference Bureau—page 107 (Palazzo dei Conservatori, Rome)

British Museum, London, and Trustees of the British Museum—pages 31, 41, 42, 92, 109

Burndy Library, Norwalk, Connecticut—pages 51, 53, 123 (from *Portraits and Lives of Illustrious Men* by André Thevet, Keruert et Chaudière, Paris, 1584)

Cabinet des Médailles, Bibliothèque Nationale, Paris—pages 72, 78, 128 *top*

The Cleveland Museum of Art—page 28 (gift of Mrs. Ralph King)

Deutsches Archäologisches Institut, Rome—pages 48–49 (Tripoli Museum), 60, 61 (photos A. von Gerkan), 68 (from "Das Symposion Ptolemaios II" by Franz Studniczka, 1914)

Egyptian Museum, Cairo—pages 128 *bottom*, 129 *bottom*

The Illustrated London News and Sketch Ltd.—pages 144–145 (from *The Illustrated London News*, November 23, 1929)

165

ACKNOWLEDGMENTS

Istanbul Arkeoloji Müzelen—frontispiece

Louvre Museum, Paris—pages 86–87, 129 *top* (photo Maurice Chuzeville), 134

The Metropolitan Museum of Art, New York—pages 39, 46 (gifts of Edward S. Harkness, 1926), 56 (gift of Darius Ogden Mills)

Erich Müller, Kassel, West Germany—page 118

The New York Public Library: Astor, Lenox and Tilden Foundations—pages 64 (from *Nouvelles Recherches Archéologiques à Begram* by J. Hackin, Mémoires de la Délégation Archéologique Française en Afghanistan, vol. XI, Presses Universitaires de France, Paris, 1954), 89 (from "The Well of Eratosthenes" by Howard Payn in *The Observatory*, vol. XXXVII, Taylor & Francis, London, 1914), 94–95 (Prints Division)

The Ny Carlsberg Glyptothek, Copenhagen—pages 33, 106

Pelizaeus Museum, Hildesheim, West Germany—pages 90, 138

Radio Times Hulton Picture Library, London—page 153

Dr. Robert L. Scranton, University of Chicago—pages 114, 115

Trustees of the Pierpont Morgan Library, New York—page 34

The picture on page 127 is from *The Book of Sun-Dials* by Mrs. Alfred Gatty, enlarged and re-edited by H. K. F. Eden and Eleanor Lloyd, George Bell and Sons, London, 1900. The pictures on pages 137 and 152 are from *Egyptian Obelisks* by Henry H. Gorringe, published by the author, New York, 1882. The picture on page 99 *bottom* is from *The Social and Economic History of the Hellenistic World* by M. Rostovtzeff, Vol. II, The Clarendon Press, Oxford, 1941.

Name Index

The history of ancient Alexandria includes a large cast of characters; therefore, this index of persons only may be particularly helpful to the reader. Suitable parenthetical identifications of the Ptolemies have been furnished, as many members of this dynasty appear in the book. Monarchs are listed according to their "official," not personal names, and all monarchs and patriarchs are identified as such. The abbreviation *passim* indicates scattered references to a subject in a sequence of pages.

INDEX

INDEX

About the Author

A Fellow of England's Royal Astronomical Society, and a former lecturer and teacher of astronomy at the American Museum—Hayden Planetarium in New York City, Kenneth Heuer is Science Editor for a leading New York publisher.

He is the author of *Men of Other Planets*, *The End of the World*, and *An Adventure in Astronomy*, published here and abroad, and has also contributed scientific articles and book reviews to national magazines and newspapers.